RED DOT PUBLICATIONS

INTRODUCTION TO ALTERNATING CURRENT AND INDUCTANCE

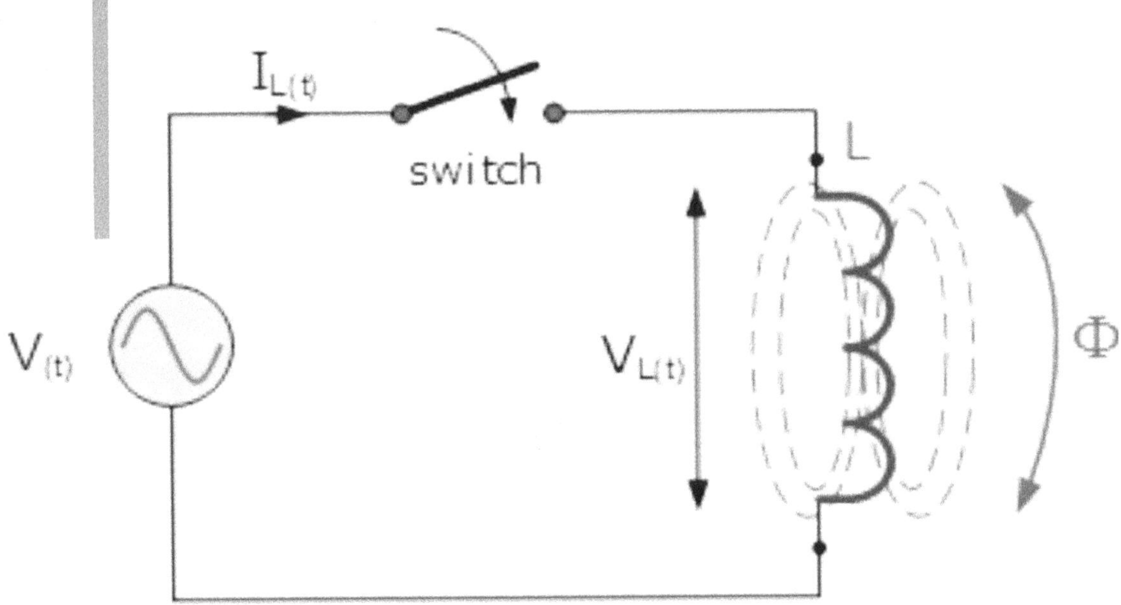

INTRODUCTION TO ALTERNATING CURRENT AND INDUCTANCE

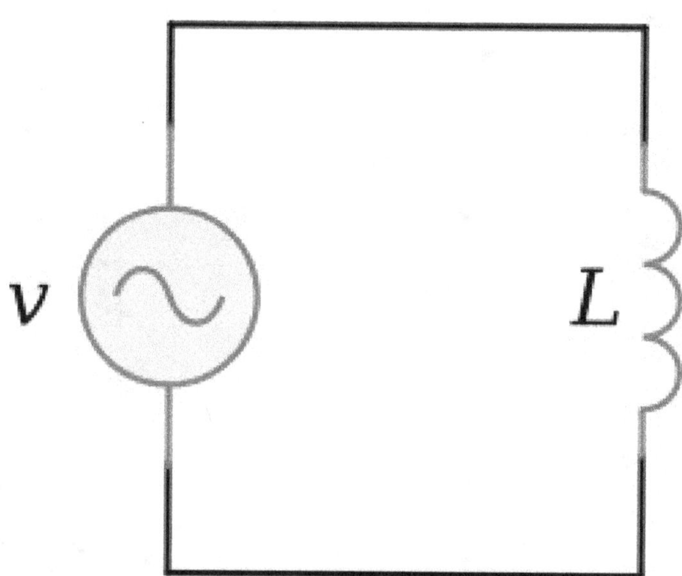

This book serves as a comprehensive guide for those seeking to understand the fundamental principles and concepts behind alternating current (AC) and inductance. Whether you are a student, an engineer, or an enthusiast eager to delve into the world of electrical circuits, this book is designed to provide you with a solid foundation in these essential topics.

Each chapter includes review questions and answers to reinforce the understanding of the material.

Chapter 1: Alternating Current

In this chapter, we embark on a journey to explore the fascinating world of alternating current. We begin by demystifying the fundamental concepts of AC, including its origins, properties, and advantages over direct current (DC). Through clear explanations and illustrative examples, we delve into AC waveforms, frequency, amplitude, and phase. Additionally, we discuss the behavior of AC circuits, including the principles of impedance, reactance, and power factor.

Chapter 2: Inductance

Building upon the knowledge gained in Chapter 1, Chapter 2 focuses on the principles of inductance and its crucial role in AC circuits. We delve into the behavior and characteristics of inductors, examining their ability to store energy in a magnetic field. Through practical examples and theoretical discussions, we explore topics such as self-inductance, mutual inductance, magnetic fields, and the effects of inductance on AC circuits. Furthermore, we discuss the applications of inductors in various electrical systems and devices.

Throughout this book, we strive to strike a balance between theoretical explanations and practical applications. Complex concepts are presented in a clear and accessible manner, allowing readers to grasp the principles underlying AC and inductance.

Whether you are an aspiring electrical engineer, a student pursuing a degree in physics, or simply an individual with a curiosity for the fascinating world of electricity, this book will provide you with a solid foundation to navigate and understand AC circuits and the principles of inductance.

Prepare to expand your knowledge and unlock the potential of these fundamental concepts that underpin modern electrical systems and technologies.

Table of Contents

ALTERNATING CURRENT...III
Can you explain how AC power can be easily regulated to a constant voltage level?..........V
What are some examples of industrial and commercial applications where high power factor loads are common?...V
How does the reduced maintenance of AC systems compared to DC systems affect their overall cost?...V
Can you explain the difference between power factor and power factor correction?.........VI
What are some advantages of using DC systems over AC systems in certain applications?...VI
How does the cost of AC systems compare to DC systems in terms of initial installation? VI
Can you provide examples of applications where DC systems are more cost-effective than AC systems?...VIII
What are some disadvantages of using DC systems compared to AC systems in certain applications?...VIII
How does power factor correction improve the efficiency of an electrical circuit?..........VIII
Disadvantages of using DC systems compared to AC systems in certain applications. .IX
The math behind AC...XI
Can you explain how impedance and admittance are related in AC circuits?..................XIII
What are some practical applications of Fourier analysis in AC systems?.....................XIII
How are Laplace transforms used to analyze AC circuits?.................................XIII
Can you explain how the transfer function is derived from the Laplace transform in AC circuits?...XV
What are some other practical applications of Fourier analysis in AC systems?...............XV
How can the location of poles and zeros in the complex plane determine the stability of an AC circuit?...XV
Deriving the Transfer Function from Laplace Transform in AC Circuits....................XVI
Practical Applications of Fourier Analysis in AC Systems...............................XVII
Stability Analysis through Pole-Zero Locations...XVII
Stability and Pole-Zero Locations:...XVIII
Inductance...XIX
Here are some key things to know about inductors:.................................XXII
Can you explain the difference between air-core inductors and iron-core inductors?..XXIII
How do impedance measurement and inductance measurement differ in testing and measuring inductors?...XXIII
What are some common design considerations when choosing a magnetic core material for an inductor?...XXIII
Air-core inductors vs iron-core inductors:.................................XXIII
Impedance measurement vs inductance measurement:.................................XXIV
Design considerations for choosing a magnetic core material:..................XXIV
Can you provide examples of high-frequency applications where air-core inductors are commonly used?...XXV
What are some other factors that can affect the efficiency of an inductor besides the core material?...XXV
Could you explain the concept of reactance in more detail and how it relates to impedance measurement?...XXV
How does the reactance of an inductor affect the overall impedance of a circuit?.......XXVII

Can you explain the relationship between inductance and reactance in more detail?. XXVII
What are some common methods for measuring the impedance of a circuit?.............XXVII
Examples for practical applications of inductance..XXIX
Can you explain how inductors are used in power supplies to store and release energy?
....XXX
What other components are commonly used in conjunction with inductors in electronic
filters?...XXX
How do inductors protect circuits from overcurrent? Can you provide more details on how
the fuse works?..XXX
    Inductors in Power Supplies:..XXX
    Other Components Used in Conjunction with Inductors in Electronic Filters:.........XXXI
    Inductors Protecting Circuits from Overcurrent:..XXXI

CHAPTER 1
CONCEPTS OF ALTERNATING CURRENT   1

CHAPTER 2   35
INDUCTANCE

ALTERNATING CURRENT

Alternating current (AC) is an electric current that periodically reverses direction, in contrast to direct current (DC) which flows in only one direction. AC is the form most commonly used for electric power distribution, as it is easy to transform and transmit over long distances without significant energy loss.

The concept of AC was first introduced by Nikola Tesla in the late 19th century, and it has since become the standard for electrical power systems worldwide. The AC waveform is created by using a sinusoidal wave, which allows for efficient transmission and distribution of power.

One of the key advantages of AC is that it can be easily transformed to higher or lower voltages using a transformer. This makes it possible to transmit power over long distances without significant energy loss, as the voltage can be stepped up or down as needed. In addition, AC is less prone to electrical shock hazards than DC, as the current is constantly changing direction and is less likely to cause a dangerous electrical path through the body.

AC is used in a wide range of applications, including household electrical outlets, industrial power systems, and even the electrical systems of many vehicles. It is a fundamental aspect of modern society, and it has revolutionized the way we live and work.

There are several advantages of AC compared to DC:

1. Easier to transmit over long distances: AC power can be transmitted over long distances with less energy loss. This is because the AC waveform has a constant magnitude, whereas the DC waveform has a constant voltage. As a result, AC power can be transmitted over longer distances without significant voltage drop.

2. Better voltage regulation: AC power can be easily regulated to a constant voltage level, which is important for many industrial and commercial applications. This is because the AC waveform can be easily filtered to remove any voltage fluctuations, resulting in a stable voltage output.

3. Reduced electromagnetic interference (EMI): AC power generates less EMI than DC power, which is important for applications where electromagnetic interference could be a problem. This is because the AC waveform has a constantly changing direction, which reduces the amount of electromagnetic radiation emitted.

4. Improved motor performance: AC motors are more efficient and have smoother operation than DC motors. This is because the AC waveform can be easily controlled to produce a sinusoidal torque, which results in smoother motor operation and improved efficiency.

5. Better power factor: AC power has a better power factor than DC power, which means that it can be used to drive loads with a high power factor. This is important for industrial and commercial applications where high power factor loads are common.

6. Reduced battery requirements: AC power does not require batteries to store energy, which is a major advantage over DC power. This is because AC power can be easily generated and transmitted over long distances without the need for batteries.

7. Improved safety: AC power is considered safer than DC power because it is less likely to cause electrical shock. This is because the AC waveform has a constantly changing direction, which reduces the risk of electrical shock.

8. Reduced maintenance: AC systems require less maintenance than DC systems, which is a major advantage in industrial and commercial applications. This is because AC systems have fewer moving parts and are less prone to wear and tear.

9. Improved efficiency: AC systems are generally more efficient than DC systems, which means that they can perform the same tasks with less energy. This is important for industrial and commercial applications where energy efficiency is a major concern.

Overall, AC power has many advantages over DC power, which is why it is widely used in industrial and commercial applications. Its ability to be easily transmitted over long distances, regulated to a constant voltage level, and used to drive loads with a high power factor make it an ideal choice for many applications.

Can you explain how AC power can be easily regulated to a constant voltage level?

What are some examples of industrial and commercial applications where high power factor loads are common?

How does the reduced maintenance of AC systems compared to DC systems affect their overall cost?

AC power can be easily regulated to a constant voltage level using a device called a transformer. A transformer is a type of electrical device that transfers electrical energy from one circuit to another through electromagnetic induction. It consists of two coils of wire, called the primary and secondary coils, that are wrapped around a common magnetic core.

When an AC voltage is applied to the primary coil, it creates a magnetic field that induces a voltage in the secondary coil. By adjusting the number of turns in the primary and secondary coils, the voltage can be stepped up or down to the desired level. This process is known as voltage transformation, and it allows AC power to be regulated to a constant voltage level.

Examples of industrial and commercial applications where high power factor loads are common include:

* Industrial motors: Many industrial motors, such as those used in pumps, fans, and conveyor systems, have a high power factor because they are designed to operate at a constant speed.

* Lighting systems: Many lighting systems, such as fluorescent and LED lighting, have a high power factor because they use electronic ballasts that regulate the flow of current to the lamps.

* HVAC systems: Heating, ventilation, and air conditioning (HVAC) systems often have high power factor loads because they use motors and compressors that operate at a constant speed.

* Power supplies: Many power supplies used in electronic devices, such as computers and televisions, have high power factor loads because they use switching regulators that regulate the output voltage.

The reduced maintenance of AC systems compared to DC systems can have a significant impact on their overall cost. AC systems have fewer moving parts than DC systems, which means that they are less prone to wear and tear. This means that AC systems require less maintenance and have a longer lifespan than DC systems.

In addition, AC systems are generally easier to repair and replace than DC systems. This is because AC systems use standardized components, such as transformers and motors, that can be easily replaced or repaired. DC systems, on the other hand, often require custom-designed components, which can be more expensive and time-consuming to replace or repair.

Overall, the reduced maintenance and repair costs of AC systems can make them a more cost-effective choice than DC systems for many industrial and commercial applications.

Can you explain the difference between power factor and power factor correction?

What are some advantages of using DC systems over AC systems in certain applications?

How does the cost of AC systems compare to DC systems in terms of initial installation?

Power factor refers to the ratio of the real power (kW) to the apparent power (kVA) in an electrical circuit. Real power is the actual power consumed by a load to perform useful work, while apparent power is the total power drawn from the source, including both real power and reactive power (kVAR). Power factor is calculated as kW / kVA.

Power factor correction (PFC) is a technique used to improve the power factor of an electrical circuit. PFC is used to reduce the amount of reactive power consumed by a load, which can help to increase the efficiency of the system and reduce the amount of energy wasted. PFC is typically achieved using a device called a power factor correction capacitor, which is connected in parallel with the load and provides a leading current to the load to offset the lagging current caused by the load's inductance.

Now, let's talk about the advantages of using DC systems over AC systems in certain applications:

1. Efficiency: DC systems are generally more efficient than AC systems because they avoid the energy losses associated with AC systems, such as the conversion losses in the transformer and the motor.

2. Cost: DC systems can be more cost-effective than AC systems in certain applications, especially for low-power systems or systems that require a high degree of control over the speed and torque of the motor.

3. Size: DC systems are often smaller and more compact than AC systems, which can be advantageous in applications where space is limited.

4. Reliability: DC systems are generally more reliable than AC systems because they have fewer moving parts and are less prone to wear and tear.

5. Control: DC systems are easier to control and regulate than AC systems, which can be advantageous in applications where precise control over the speed and torque of the motor is required.

Finally, let's talk about the cost of AC systems compared to DC systems in terms of initial installation:

1. AC systems typically have a higher initial cost than DC systems because they require more components, such as transformers and switchgear, to step up and step down the voltage.

2. DC systems, on the other hand, require fewer components and can be simpler to install, which can result in lower initial costs.

3. However, it's important to note that the cost of AC systems can be offset by their higher efficiency and longer lifespan, which can result in lower operating costs over time.

4. The choice between AC and DC systems ultimately depends on the specific requirements of the application and the trade-offs between initial cost, efficiency, and reliability.

Can you provide examples of applications where DC systems are more cost-effective than AC systems?

What are some disadvantages of using DC systems compared to AC systems in certain applications?

How does power factor correction improve the efficiency of an electrical circuit?

Examples of applications where DC systems are more cost-effective than AC systems:

1. Low-power applications: DC systems are often more cost-effective than AC systems for low-power applications, such as small appliances, electronic devices, and lighting systems.

2. Battery-powered systems: DC systems are commonly used in battery-powered systems, such as electric vehicles, mobile devices, and renewable energy systems, because they can operate directly from the battery without the need for an inverter.

3. Industrial control systems: DC systems are often used in industrial control systems, such as motor control, lighting control, and process control, because they can provide a high degree of control over the speed and torque of the motor.

4. Electric vehicles: DC systems are commonly used in electric vehicles because they can provide a high degree of control over the speed and torque of the motor, and they can operate directly from the battery without the need for an inverter.

Disadvantages of using DC systems compared to AC systems in certain applications

1. Limited power handling capacity: DC systems have a limited power handling capacity compared to AC systems, which can make them less suitable for high-power applications.

2. Limited flexibility: DC systems are less flexible than AC systems because they require a fixed voltage and current to operate, which can limit their ability to adapt to changing load conditions.

3. Limited distance transmission: DC systems are less suitable for long-distance transmission because they lose voltage and current over distance, which can result in reduced efficiency and power output.

4. Higher cost per watt: DC systems can be more expensive per watt than AC systems, which can make them less cost-effective for high-power applications.

Power factor correction (PFC) improves the efficiency of an electrical circuit by reducing the amount of reactive power consumed by the load. Reactive power is the portion of the total power consumed by a load that is not used to perform useful work, but instead is used to sustain the magnetic field of the load. PFC reduces the amount of reactive power consumed by the load by providing a leading current to the load, which offsets the lagging current caused by the load's inductance. This results in a higher power factor, which means that a higher percentage of the total power consumed by the load is used to perform useful work.

PFC can be achieved using a variety of techniques, including passive PFC, active PFC, and hybrid PFC. Passive PFC uses a network of capacitors and inductors to provide a leading current to the load, while active PFC uses an active device, such as an insulated gate bipolar transistor (IGBT), to actively control the flow of current to the load. Hybrid PFC combines passive and active PFC techniques to achieve a balance between cost and performance.

The math behind AC

The math behind AC (Alternating Current) systems is based on the principles of electromagnetism and the behavior of sinusoidal waves. Here are some key concepts and equations that describe the math of AC systems:

1. Sinusoidal Waves: AC systems use sinusoidal waves to represent the alternating current and voltage. A sinusoidal wave is a continuous wave that has a constant amplitude and frequency, and can be represented by the equation:

$$y(t) = A \sin(2\pi f t + \varphi)$$

where $y(t)$ is the waveform, A is the amplitude, f is the frequency, and φ is the phase angle.

2. Complex Exponentials: To analyze AC circuits, we often use complex exponentials, which are mathematical expressions that describe the behavior of sinusoidal waves in the complex plane. Complex exponentials can be represented by the equation:

$$e^{(j\omega t)} = \cos(\omega t) + j \sin(\omega t)$$

where j is the imaginary unit ($j = \sqrt{(-1)}$), ω is the angular frequency (related to the frequency f by $\omega = 2\pi f$), and t is time.

3. Impedance: Impedance is a measure of the total opposition to the flow of an alternating current in an electrical circuit. It is a complex quantity, represented by the symbol Z, and can be calculated using the equation:

$$Z = R + jX$$

where R is the resistance and X is the reactance of the circuit.

4. Admittance: Admittance is a measure of the ability of a circuit to store energy in an electric field. It is the inverse of impedance, and can be represented by the symbol Y. Admittance can be calculated using the equation:

$$Y = 1/Z$$

5. Power Factor: The power factor of an AC circuit is a measure of how effectively the circuit is using energy. It is defined as the ratio of real power (kW) to apparent power (kVA), and can be calculated using the equation:

$$power\ factor = kW / kVA$$

6. AC Circuit Analysis: To analyze AC circuits, we use techniques such as phasors, which are mathematical representations of sinusoidal waves in the complex plane. Phasors can be used to represent voltage, current, and power in an AC circuit, and can be manipulated using trigonometric identities and equations.

7. Fourier Analysis: Fourier analysis is a method for decomposing a complex waveform into its component frequencies. It is used to analyze the frequency content of AC signals, and can be useful for filtering and modulation applications. Fourier analysis can be performed using the Fourier transform equation:

$$X(f) = \int x(t)e^{\wedge}(-j\omega t)dt$$

where X(f) is the Fourier transform of the waveform x(t), and ω is the angular frequency.

8. Laplace Transforms: Laplace transforms are a powerful tool for analyzing AC circuits. They allow us to convert differential equations into algebraic equations, which can be easier to solve. The Laplace transform of a function f(t) is represented by the symbol F(s) and can be calculated using the equation:

$$F(s) = \int f(t)e^{\wedge}(-st)dt$$

where s is a complex variable.

These are just a few of the key concepts and equations that describe the math behind AC systems. Understanding these concepts is crucial for designing, analyzing, and optimizing AC circuits, which are used in a wide range of applications, from power generation and distribution to audio and radio communication systems.

Can you explain how impedance and admittance are related in AC circuits?

What are some practical applications of Fourier analysis in AC systems?

How are Laplace transforms used to analyze AC circuits?

Impedance and admittance are related in AC circuits in that they are both measures of the opposition to the flow of alternating current in a circuit. Impedance is a measure of the total opposition to the flow of current, including both resistance and reactance, while admittance is a measure of the ability of a circuit to store energy in an electric field.

In AC circuits, impedance and admittance are related by the following equation:

$$Z = 1/Y$$

where Z is the impedance and Y is the admittance. This equation states that the impedance of a circuit is inversely proportional to its admittance. In other words, if a circuit has a high impedance, it has a low admittance, and vice versa.

Fourier analysis is a powerful tool for analyzing AC systems because it allows us to decompose a complex waveform into its component frequencies. This can be useful for a variety of practical applications, such as:

1. Filtering: Fourier analysis can be used to design filters that reject all frequencies except for a specific range of interest. For example, in audio applications, a filter might be designed to reject all frequencies except for a specific range of frequencies corresponding to a particular musical note.

2. Modulation: Fourier analysis can be used to analyze the frequency content of a modulated signal, which is useful for demodulating the signal and extracting the original information.

3. Image processing: Fourier analysis can be used to analyze and process images, which is useful for applications such as image compression and enhancement.

4. Power system analysis: Fourier analysis can be used to analyze the frequency content of power systems, which is useful for designing and optimizing power grids.

Laplace transforms are used to analyze AC circuits by converting differential equations into algebraic equations. The Laplace transform of a function f(t) is represented by the symbol F(s) and is defined as:

$$F(s) = \int f(t)e^{\wedge}(-st)dt$$

where s is a complex variable. The Laplace transform can be used to analyze AC circuits in several ways, including:

1. Transfer function analysis: The transfer function of an AC circuit is a mathematical representation of the relationship between the input and output voltages or currents. The transfer function can be obtained by taking the Laplace transform of the circuit equations and can be used to analyze the behavior of the circuit.

2. Frequency response analysis: The frequency response of an AC circuit can be analyzed by taking the Laplace transform of the circuit equations and plotting the resulting transfer function as a function of frequency. This can help to identify the frequency response of the circuit and to design filters or other frequency-selective components.

3. Stability analysis: The stability of an AC circuit can be analyzed by taking the Laplace transform of the circuit equations and examining the poles and zeros of the resulting transfer function. The poles of the transfer function correspond to the roots of the equation, and the zeros correspond to the zeros of the equation. The stability of the circuit can be determined by examining the location of the poles and zeros in the complex plane.

Can you explain how the transfer function is derived from the Laplace transform in AC circuits?

What are some other practical applications of Fourier analysis in AC systems?

How can the location of poles and zeros in the complex plane determine the stability of an AC circuit?

Deriving the Transfer Function from Laplace Transform in AC Circuits

Modeling the Circuit:

Imagine a simple AC circuit like the one below, containing a resistor (R), capacitor (C), and inductor (L) connected in series. We want to analyze how the output voltage (Vout) behaves when an input voltage (Vin) is applied.

To use the Laplace transform, we first replace each circuit element with its corresponding Laplace domain impedance:

- Resistor: R (remains the same)
- Capacitor: 1/sC
- Inductor: sL

Applying the Laplace Transform:

Now, we take the Laplace transform of both the input and output voltages:

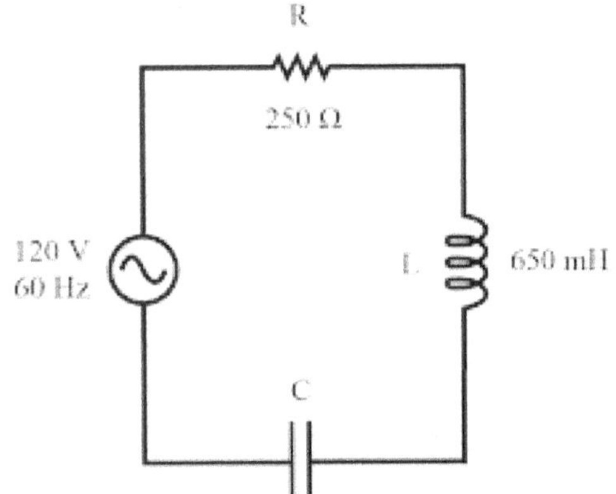

- Vin(s) = Laplace transform of Vin(t)
- Vout(s) = Laplace transform of Vout(t)

Rearranging and Simplifying:

Divide the output transform by the input transform to eliminate common factors:

```
H(s) = Vout(s) / Vin(s)
```

This expression, H(s), is the transfer function. It relates the output voltage in the frequency domain (Vout(s)) to the input voltage (Vin(s)) for any frequency.

Example:

For the simple circuit above, the transfer function can be derived as:

```
H(s) = R / (Rs + sL + 1/sC)
```

This equation tells us how the amplitude and phase of the output voltage will change relative to the input voltage at different frequencies.

Practical Applications of Fourier Analysis in AC Systems

Fourier analysis decomposes complex signals into their constituent frequencies, revealing the spectral content. This has various practical applications:

1. Harmonic Analysis:

Imagine a noisy power signal. By applying Fourier analysis, we can identify the specific frequencies of the noise components, allowing us to design filters to remove them and obtain a cleaner signal.

2. Frequency Response Analysis:

By studying the amplitude and phase shift of a system's output for different input frequencies, we can:

- Predict how the circuit will behave at different frequencies.
- Design amplifiers to boost specific frequency ranges.
- Optimize communication systems for efficient data transmission.

3. Transient Analysis:

Fourier analysis helps analyze how a system responds to sudden changes in the input signal, such as power surges or switch-on transients. This is crucial for understanding the behavior of power electronics and control systems.

Stability Analysis through Pole-Zero Locations

The stability of an AC circuit depends on the location of its poles and zeros in the complex plane:

Poles:

- Represent frequencies where the denominator of the transfer function becomes zero (infinite gain).
- If any pole has a positive real part, the system is unstable and will oscillate uncontrollably.

Zeros:

- Represent frequencies where the numerator of the transfer function becomes zero (zero gain).
- They don't directly affect stability but influence the system's gain and frequency response.

-

Stability and Pole-Zero Locations:

A stable system must have all its poles with **negative real parts**. The closer the poles are to the imaginary axis, the slower the system's response and the higher the damping factor (indicating less oscillation). Conversely, poles closer to the positive real axis indicate a less damped response and potential instability.

Understanding pole-zero locations is essential for designing stable amplifiers, filters, and control systems in AC circuits.

Inductance

Inductance is a measure of how much a conductor resists changes in current flow. It is a property of the conductor itself, and is related to the physical structure of the conductor, such as the number of turns in a coil or the cross-sectional area of a wire.

Inductance is caused by the magnetic field that is generated by the current flowing through a conductor. When the current flowing through a conductor changes, the magnetic field around the conductor also changes. This change in the magnetic field induces an electromotive force (EMF) in the conductor, which opposes the change in current. This opposition to changes in current flow is known as inductance.

The unit of inductance is the henry (H), and it is defined as the ratio of the inductive voltage to the rate of change of the current. In other words, it is a measure of how much the inductor resists changes in current flow.

Inductors are commonly used in electronic circuits for a variety of purposes, including:

* Filtering: Inductors can be used to block high-frequency signals and allow low-frequency signals to pass through, making them useful for filtering applications.

* Energy storage: Inductors can store energy in the form of a magnetic field, and release it when needed.

* Impedance matching: Inductors can be used to match the impedance of a circuit to a load, improving the efficiency of the system.

The behavior of an inductor can be described using the following equations:

* The inductance of an inductor is given by the equation: L = N * μ_o * A / L

Where:

* L is the inductance in henries (H)

* N is the number of turns of the coil

* μ_0 is the magnetic permeability of the core material (if present)

* A is the cross-sectional area of the coil

* L is the length of the coil

* The current flowing through an inductor is given by the equation: I = V / (R + jX)

Where:

* I is the current in amperes (A)

* V is the voltage across the inductor in volts (V)

* R is the resistance of the inductor in ohms (Ω)

* X is the reactance of the inductor in ohms (Ω)

* j is the imaginary unit (j = $\sqrt{(-1)}$)

* The impedance of an inductor is given by the equation: Z = R + jX

Where:

* Z is the impedance in ohms (Ω)

* R is the resistance of the inductor in ohms (Ω)

* X is the reactance of the inductor in ohms (Ω)

* j is the imaginary unit (j = $\sqrt{(-1)}$)

Inductors are an important component in many electronic circuits, and understanding their behavior is crucial for designing and analyzing these circuits.

The concept of inductance is based on the idea that a magnetic field can store energy. When a current flows through a conductor, such as a wire, it generates a magnetic field around the conductor. The strength of the magnetic field depends on the amount of current flowing through the conductor.

If the current flowing through the conductor changes, the magnetic field around the conductor also changes. This change in the magnetic field induces an electromotive force (EMF) in the conductor, which opposes the change in current. This opposition to changes in current flow is known as inductance.

The amount of inductance in a conductor depends on several factors, including the number of turns in a coil, the cross-sectional area of the coil, and the magnetic permeability of the core material (if present). The unit of inductance is the henry (H), and it is defined as the ratio of the inductive voltage to the rate of change of the current.

Inductors are commonly used in electronic circuits for a variety of purposes, including:

* Filtering: Inductors can be used to block high-frequency signals and allow low-frequency signals to pass through, making them useful for filtering applications.

* Energy storage: Inductors can store energy in the form of a magnetic field, and release it when needed.

* Impedance matching: Inductors can be used to match the impedance of a circuit to a load, improving the efficiency of the system.

* Oscillation: Inductors can be used to create oscillations in a circuit, making them useful for creating clock signals in digital circuits.

Inductors are also used in many other applications, such as power supplies, transformers, and motors.

Inductors are passive electrical components that store energy in a magnetic field. They consist of a coil of wire wrapped around a core material, and they are commonly used in electronic circuits for a variety of purposes.

Here are some key things to know about inductors:

1. Inductance: The ability of an inductor to store energy in a magnetic field is measured in henries (H). The higher the inductance, the more energy the inductor can store.

2. Current: Inductors oppose changes in current flow. When the current through an inductor changes, the magnetic field around the coil also changes, inducing an electromotive force (EMF) that opposes the change in current.

3. Voltage: The voltage across an inductor is proportional to the rate of change of the current flowing through it. This is known as Faraday's law of induction.

4. Energy storage: Inductors can store energy in the form of a magnetic field. The energy stored in an inductor is proportional to the square of the current flowing through it.

5. Magnetic core: The magnetic core of an inductor is the material inside the coil that provides a path for the magnetic field to flow. The magnetic core can be made of a variety of materials, including iron, nickel, and ferrite.

6. Types of inductors: There are several types of inductors, including air-core inductors, iron-core inductors, and toroidal inductors. Each type has its own characteristics and is used in different applications.

7. Applications: Inductors are used in a wide range of applications, including power supplies, transformers, motors, and electronic filters. They are also used in many consumer electronics, such as computers, televisions, and smartphones.

8. Design considerations: When designing an inductor, there are several factors to consider, including the desired inductance, the current rating, the magnetic core material, and the physical dimensions.

9. Mathematical modeling: Inductors can be modeled mathematically using equations that describe the behavior of the magnetic field and the current flowing through the coil. These equations can be used to design and analyze inductors for specific applications.

10. Testing and measurement: Inductors can be tested and measured using a variety of techniques, including impedance measurement, inductance measurement, and magnetic field measurement.

Can you explain the difference between air-core inductors and iron-core inductors?

How do impedance measurement and inductance measurement differ in testing and measuring inductors?

What are some common design considerations when choosing a magnetic core material for an inductor?

Air-core inductors vs iron-core inductors:

Air-core inductors are inductors that do not have a magnetic core material. Instead, the coil is wound on a non-magnetic material, such as plastic or ceramic. Air-core inductors have a lower inductance than iron-core inductors, but they are smaller and lighter. They are commonly used in high-frequency applications, such as in radio frequency (RF) circuits and switched-mode power supplies.

Iron-core inductors, on the other hand, have a magnetic core material, such as iron or ferrite, inside the coil. The magnetic core material increases the inductance of the coil, making it more efficient. Iron-core inductors are commonly used in low-frequency applications, such as in power supplies and motor control circuits.

Impedance measurement vs inductance measurement:

Impedance measurement is a measure of the total opposition to the flow of an alternating current (AC) in an electrical circuit. It includes both resistance and reactance. Inductance measurement, on the other hand, is a measure of the amount of magnetic field stored in an inductor.

In testing and measuring inductors, impedance measurement is typically used to determine the impedance of the inductor, which can be used to determine the current-carrying capacity of the inductor. Inductance measurement, on the other hand, is used to determine the inductance of the inductor, which can be used to determine the energy storage capacity of the inductor.

Design considerations for choosing a magnetic core material:

When choosing a magnetic core material for an inductor, there are several design considerations to keep in mind. Some of these considerations include:

1. Magnetic permeability: The magnetic permeability of the core material determines the efficiency of the inductor. Materials with high magnetic permeability, such as ferrite, are preferred for inductor cores.

2. Saturation point: The saturation point of the core material determines the maximum magnetic field that can be stored in the inductor. Materials with a higher saturation point, such as iron, are preferred for high-power inductors.

3. Losses: The losses in the core material, such as hysteresis losses and eddy current losses, can reduce the efficiency of the inductor. Materials with low losses, such as ferrite, are preferred for inductor cores.

4. Cost: The cost of the core material is an important consideration in the design of inductors. Materials such as ferrite and iron are relatively inexpensive, making them popular choices for inductor cores.

5. Size and shape: The size and shape of the core material can affect the performance of the inductor. For example, a toroidal shape can provide a higher inductance than a rectangular shape.

6. Operating temperature: The operating temperature of the inductor can affect the performance of the core material. Materials such as ferrite and iron have a high Curie temperature, making them suitable for high-temperature applications.

7. Magnetic field strength: The magnetic field strength of the inductor can affect the performance of the core material. Materials with a high magnetic field strength, such as neodymium, are preferred for high-power inductors.

These are some of the key design considerations when choosing a magnetic core material for an inductor. The specific requirements of the application will determine the best material choice.

Can you provide examples of high-frequency applications where air-core inductors are commonly used?

What are some other factors that can affect the efficiency of an inductor besides the core material?

Could you explain the concept of reactance in more detail and how it relates to impedance measurement?

Examples of high-frequency applications where air-core inductors are commonly used include:

* Radio frequency (RF) circuits: Air-core inductors are often used in RF circuits because they have low losses and can handle high frequencies. They are commonly used in filters, tuners, and oscillators.

* Switched-mode power supplies: Air-core inductors are used in switched-mode power supplies because they can handle high frequencies and have low losses. They are used to store energy and filter out unwanted frequencies.

* Audio equipment: Air-core inductors are used in audio equipment, such as audio filters and equalizers, because they can handle high frequencies and have low losses.

Other factors that can affect the efficiency of an inductor besides the core material include:

* Winding configuration: The way the coil is wound can affect the efficiency of the inductor. For example, a spiral winding can be more efficient than a straight winding.

* Coil size and shape: The size and shape of the coil can affect the efficiency of the inductor. A larger coil with a larger cross-sectional area can be more efficient than a smaller coil.

* Number of turns: The number of turns in the coil can affect the efficiency of the inductor. More turns can increase the inductance, but can also increase the losses.

* Insulation: The insulation material used in the coil can affect the efficiency of the inductor. Some insulation materials, such as plastic, can have high dielectric losses, which can reduce the efficiency of the inductor.

Reactance is the opposition to current flow in an electrical circuit due to the presence of inductive and capacitive elements. It is measured in ohms and is the imaginary part of the impedance of a circuit.

Impedance measurement is a measure of the total opposition to the flow of an alternating current (AC) in an electrical circuit. It includes both resistance and reactance. Reactance is the opposition to current flow due to the presence of inductive and capacitive elements, while resistance is the opposition to current flow due to the resistance of the conductor.

In the case of an inductor, the reactance is due to the magnetic field that is generated by the current flowing through the coil. The magnetic field stores energy and opposes changes in current flow. The reactance of an inductor is given by the equation:

$$X_L = 2\pi f L$$

Where X_L is the reactance, f is the frequency of the current, and L is the inductance.

Impedance measurement is important in circuit design because it determines the current flowing through a circuit. By measuring the impedance of a circuit, engineers can determine the current-carrying capacity of the circuit and select components that can handle the current.

How does the reactance of an inductor affect the overall impedance of a circuit?

Can you explain the relationship between inductance and reactance in more detail?

What are some common methods for measuring the impedance of a circuit?

The reactance of an inductor affects the overall impedance of a circuit by opposing the current flow in the circuit. The reactance of an inductor is a measure of how much the inductor resists changes in current flow. When the current flowing through an inductor changes, the magnetic field around the inductor also changes, and this change in magnetic field induces a voltage across the inductor that opposes the change in current. This opposition to changes in current flow is known as reactance.

The reactance of an inductor is given by the equation:

$$X_L = 2\pi f L$$

Where XL is the reactance, f is the frequency of the current, and L is the inductance.

The reactance of an inductor is a complex quantity, meaning that it has both magnitude and phase. The magnitude of the reactance is determined by the inductance and the frequency of the current, while the phase is determined by the angle between the current and the voltage across the inductor.

The relationship between inductance and reactance is that reactance is a measure of the opposition to current flow in an inductor, while inductance is a measure of the amount of magnetic field that is generated by the current flowing through the inductor. The greater the inductance, the greater the reactance, and the more the inductor resists changes in current flow.

There are several common methods for measuring the impedance of a circuit, including:

1. Ohm's law: This method involves measuring the voltage across a resistor and the current flowing through it, and then using Ohm's law to calculate the impedance.

2. Impedance measurement using a vector impedance analyzer: This method involves using a vector impedance analyzer to measure the magnitude and phase of the impedance of a circuit.

3. Impedance measurement using a network analyzer: This method involves using a network analyzer to measure the magnitude and phase of the impedance of a circuit.

4. Impedance measurement using a spectrum analyzer: This method involves using a spectrum analyzer to measure the magnitude and phase of the impedance of a circuit over a range of frequencies.

5. Impedance measurement using a lock-in amplifier: This method involves using a lock-in amplifier to measure the magnitude and phase of the impedance of a circuit by applying a small AC signal to the circuit and measuring the resulting voltage and current.

Examples for practical applications of inductance

Inductance has many practical applications in various fields, including electronics, electrical engineering, and physics. Here are some examples:

1. Filtering: Inductors are used in electronic filters to block certain frequencies and allow others to pass through. For example, a low-pass filter can be made by connecting an inductor in series with a resistor and a capacitor.

2. Energy storage: Inductors can store energy in the form of a magnetic field. This property is used in many applications, such as in power supplies, where an inductor is used to store energy and release it to the circuit as needed.

3. Impedance matching: Inductors are used to match the impedance of a circuit to a load, ensuring maximum power transfer. For example, a transformer can be used to step up or step down the voltage of a circuit to match the impedance of a load.

4. Oscillation: Inductors can be used to create oscillations in a circuit. For example, a tank circuit, which consists of an inductor and a capacitor, can be used to create a sinusoidal oscillation.

5. Measurement: Inductors can be used to measure the frequency of a circuit. For example, a resonant circuit, which consists of an inductor and a capacitor, can be used to measure the frequency of a signal.

6. Safety: Inductors can be used to protect circuits from overcurrent. For example, a fuse can be made by connecting an inductor in series with a circuit. When the current exceeds a certain limit, the inductor opposes the change in current, preventing damage to the circuit.

7. Medical applications: Inductors are used in some medical devices, such as MRI machines, which use powerful magnetic fields to create images of the body.

8. Automotive applications: Inductors are used in some automotive applications, such as in the ignition system of a car, where an inductor is used to generate a high voltage to spark the fuel in the engine.

9. Aerospace applications: Inductors are used in some aerospace applications, such as in the navigation systems of aircraft, where an inductor is used to

generate a magnetic field that can be used to determine the direction of the aircraft.

10. Power transmission: Inductors are used in power transmission systems, such as in power lines, where they are used to reduce energy losses and improve the efficiency of the system.

These are just a few examples of the many practical applications of inductance. Inductors are versatile components that can be used in a wide range of circuits and systems, and their properties make them useful in many different fields.

Can you explain how inductors are used in power supplies to store and release energy?

What other components are commonly used in conjunction with inductors in electronic filters?

How do inductors protect circuits from overcurrent? Can you provide more details on how the fuse works?

Inductors in Power Supplies:

Inductors are commonly used in power supplies to store and release energy. The inductor acts as a reservoir for magnetic energy, which can be used to supply power to a circuit when the voltage source is not available. When the voltage source is present, the inductor stores energy in the form of a magnetic field. When the voltage source is removed, the inductor releases the stored energy to the circuit, allowing it to continue operating.

The inductor is typically used in conjunction with a capacitor and a diode in a power supply circuit. The capacitor is used to smooth out the output voltage, while the diode is used to rectify the AC input voltage and provide a DC output voltage. The inductor is used to store energy and release it as needed to maintain a stable output voltage.

Other Components Used in Conjunction with Inductors in Electronic Filters:

Inductors are commonly used in electronic filters in conjunction with capacitors, resistors, and active devices such as op-amps. The inductor and capacitor are used to form a resonant circuit, which filters out unwanted frequencies and allows desired frequencies to pass through. The resistor is used to set the gain of the filter, while the op-amp is used to amplify the filtered signal.

Inductors Protecting Circuits from Overcurrent:

Inductors can protect circuits from overcurrent by opposing changes in current flow. When the current flowing through an inductor changes, the inductor generates a back-electromotive force (EMF) that opposes the change in current. This property of inductors is known as Lenz's law.

A fuse is a type of inductor that is specifically designed to protect circuits from overcurrent. When the current flowing through a fuse exceeds a certain limit, the fuse heats up and melts, breaking the circuit and preventing damage to the rest of the circuit. The fuse is designed to have a specific current rating, which is the maximum current that it can handle without melting.

In addition to fuses, other types of inductors can also be used to protect circuits from overcurrent. For example, a coil or a solenoid can be used to oppose changes in current flow and prevent overcurrent from damaging the circuit.

CHAPTER 1

CONCEPTS OF ALTERNATING CURRENT

LEARNING OBJECTIVES

Upon completion of this chapter you will be able to:

1. State the differences between ac and dc voltage and current.

2. State the advantages of ac power transmission over dc power transmission.

3. State the "left-hand rule" for a conductor.

4. State the relationship between current and magnetism.

5. State the methods by which ac power can be generated.

6. State the relationship between frequency, period, time, and wavelength.

7. Compute peak-to-peak, instantaneous, effective, and average values of voltage and current.

8. Compute the phase difference between sine waves.

CONCEPTS OF ALTERNATING CURRENT

All of your study thus far has been with direct current (dc), that is, current which does not change direction. However, as you saw in module 1 and will see later in this module, a coil rotating in a magnetic field actually generates a current which regularly changes direction. This current is called ALTERNATING CURRENT or ac.

AC AND DC

Alternating current is current which constantly changes in amplitude, and which reverses direction at regular intervals. You learned previously that direct current flows only in one direction, and that the amplitude of current is determined by the number of electrons flowing past a point in a circuit in one second. If, for example, a coulomb of electrons moves past a point in a wire in one second and all of the electrons are moving in the same direction, the amplitude of direct current in the wire is one ampere. Similarly, if half a coulomb of electrons moves in one direction past a point in the wire in half a second, then reverses direction and moves past the same point in the opposite direction during the next half-second, a total of one coulomb of electrons passes the point in one second. The amplitude of the alternating current is one ampere. The preceding comparison of dc and ac as illustrated. Notice that one white arrow plus one striped arrow comprise one coulomb.

COMPARING DC & AC CURRENT FLOW IN A WIRE

Q1. Define direct current.

Q2. Define alternating current.

DISADVANTAGES OF DC COMPARED TO AC

When commercial use of electricity became wide-spread in the United States, certain disadvantages in using direct current in the home became apparent. If a commercial direct-current system is used, the voltage must be generated at the level (amplitude or value) required by the load. To properly light a 240-volt lamp, for example, the dc generator must deliver 240 volts. If a 120-volt lamp is to be supplied power from the 240-volt generator, a resistor or another 120-volt lamp must be placed in series with the 120-volt lamp to drop the extra 120 volts. When the resistor is used to reduce the voltage, an amount of power equal to that consumed by the lamp is wasted.

Another disadvantage of the direct-current system becomes evident when the direct current (I) from the generating station must be transmitted a long distance over wires to the consumer. When this happens, a large amount of power is lost due to the resistance (R) of the wire. The power loss is equal to I^2R. However, this loss can be greatly reduced if the power is transmitted over the lines at a very high voltage level and a low current level. This is not a practical solution to the power loss in the dc system since the load would then have to be operated at a dangerously high voltage. Because of the disadvantages related to transmitting and using direct current, practically all modern commercial electric power companies generate and distribute alternating current (ac).

Unlike direct voltages, alternating voltages can be stepped up or down in amplitude by a device called a TRANSFORMER. (The transformer will be explained later in this module.) Use of the transformer permits efficient transmission of electrical power over long-distance lines. At the electrical power station, the transformer output power is at high voltage and low current levels. At the consumer end of the transmission lines, the voltage is stepped down by a transformer to the value required by the load. Due to its inherent advantages and versatility, alternating current has replaced direct current in all but a few commercial power distribution systems.

Q3. *What is a disadvantage of a direct-current system with respect to supply voltage?*

Q4. *What disadvantage of a direct current is due to the resistance of the transmission wires?*

Q5. *What kind of electrical current is used in most modern power distribution systems?*

VOLTAGE WAVEFORMS

You now know that there are two types of current and voltage, that is, direct current and voltage and alternating current and voltage. If a graph is constructed showing the amplitude of a dc voltage across the terminals of a battery with respect to time, it will appear in figure 1-1 view A. The dc voltage is shown to have a constant amplitude. Some voltages go through periodic changes in amplitude like those shown in figure 1-1 view B. The pattern which results when these changes in amplitude with respect to time are plotted on graph paper is known as a WAVEFORM. Figure 1-1 view B shows some of the common electrical waveforms. Of those illustrated, the sine wave will be dealt with most often.

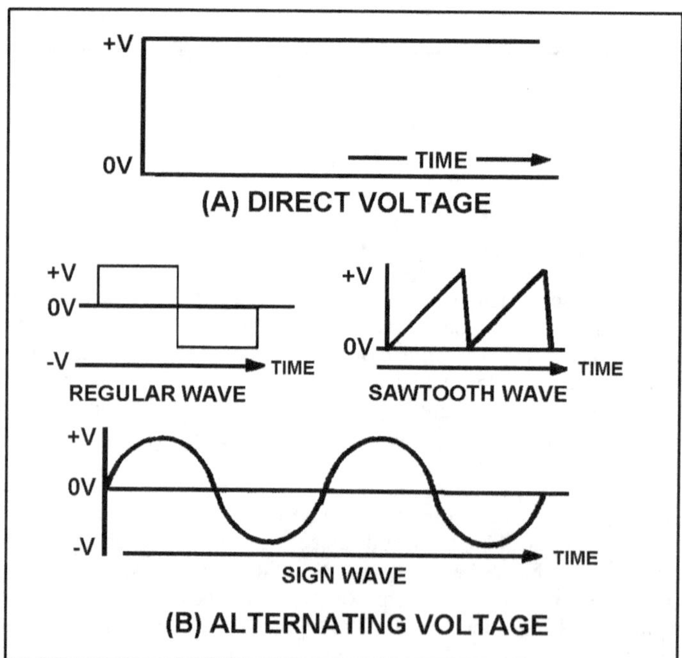

Figure 1-1.—Voltage waveforms: (A) Direct voltage; (B) Alternating voltage.

ELECTROMAGNETISM

The sine wave illustrated in figure 1-1 view B is a plot of a current which changes amplitude and direction. Although there are several ways of producing this current, the method based on the principles of electromagnetic induction is by far the easiest and most common method in use.

The fundamental theories concerning simple magnets and magnetism were discussed in Module 1, but how magnetism can be used to produce electricity was only briefly mentioned. This module will give you a more in-depth study of magnetism. The main points that will be explained are how magnetism is affected by an electric current and, conversely, how electricity is affected by magnetism. This general subject area is most often referred to as ELECTROMAGNETISM. To properly understand electricity you must first become familiar with the relationships between magnetism and electricity. For example, you must know that:

- An electric current always produces some form of magnetism.

- The most commonly used means for producing or using electricity involves magnetism.

- The peculiar behavior of electricity under certain conditions is caused by magnetic influences.

MAGNETIC FIELDS

In 1819 Hans Christian Oersted, a Danish physicist, found that a definite relationship exists between magnetism and electricity. He discovered that an electric current is always accompanied by certain magnetic effects and that these effects obey definite laws.

MAGNETIC FIELD AROUND A CURRENT-CARRYING CONDUCTOR

If a compass is placed in the vicinity of a current-carrying conductor, the compass needle will align itself at right angles to the conductor, thus indicating the presence of a magnetic force. You can demonstrate the presence of this force by using the arrangement illustrated in figure 1-2. In both (A) and (B) of the figure, current flows in a vertical conductor through a horizontal piece of cardboard. You can determine the direction of the magnetic force produced by the current by placing a compass at various points on the cardboard and noting the compass needle deflection. The direction of the magnetic force is assumed to be the direction in which the north pole of the compass points.

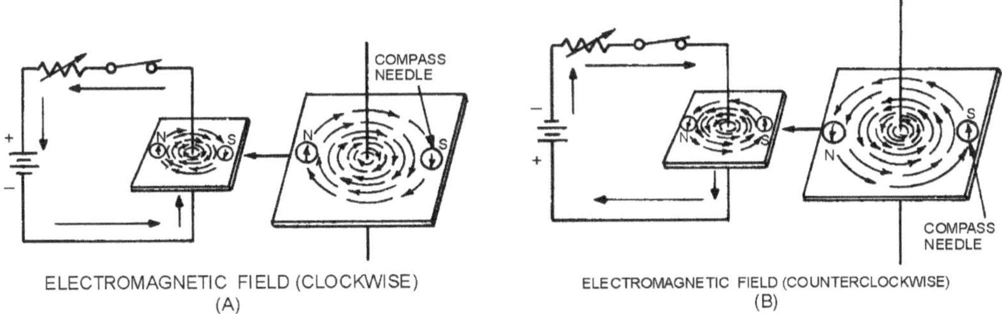

Figure 1-2.—Magnetic field around a current-carrying conductor.

In figure 1-2 (A), the needle deflections show that a magnetic field exists in circular form around the conductor. When the current flows upward (see figure 1-2(A)), the direction of the field is clockwise, as viewed from the top. However, if you reverse the polarity of the battery so that the current flows downward (see figure 1-2(B)), the direction of the field is counterclockwise.

The relation between the direction of the magnetic lines of force around a conductor and the direction of electron current flow in the conductor may be determined by means of the LEFT-HAND RULE FOR A CONDUCTOR: if you grasp the conductor in your left hand with the thumb extended in the direction of the electron flow (current) ($-$ to $+$), your fingers will point in the direction of the magnetic lines of force. Now apply this rule to figure 1-2. Note that your fingers point in the direction that the north pole of the compass points when it is placed in the magnetic field surrounding the wire.

An arrow is generally used in electrical diagrams to denote the direction of current in a length of wire (see figure 1-3(A)). Where a cross section of a wire is shown, an end view of the arrow is used. A cross-sectional view of a conductor that is carrying current toward the observer is illustrated in figure 1-3(B). Notice that the direction of current is indicated by a dot, representing the head of the arrow. A conductor that is carrying current away from the observer is illustrated in figure 1-3(C). Note that the direction of current is indicated by a cross, representing the tail of the arrow. Also note that the magnetic field around a current-carrying conductor is perpendicular to the conductor, and that the magnetic lines of force are equal along all parts of the conductor.

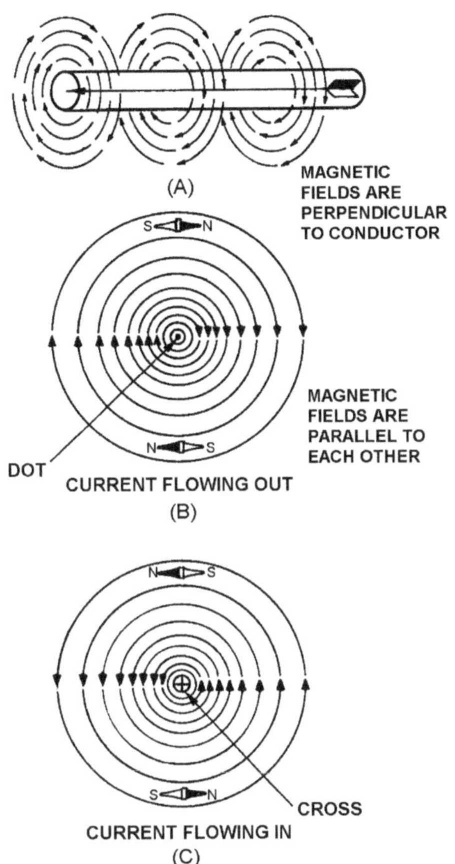

Figure 1-3.—Magnetic field around a current-carrying conductor, detailed view.

When two adjacent parallel conductors are carrying current in the same direction, the magnetic lines of force combine and increase the strength of the field around the conductors, as shown in figure 1-4(A). Two parallel conductors carrying currents in opposite directions are shown in figure 1-4(B). Note that the field around one conductor is opposite in direction to the field around the other conductor. The resulting lines of force oppose each other in the space between the wires, thus deforming the field around each conductor. This means that if two parallel and adjacent conductors are carrying currents in the same direction, the fields about the two conductors aid each other. Conversely, if the two conductors are carrying currents in opposite directions, the fields about the conductors repel each other.

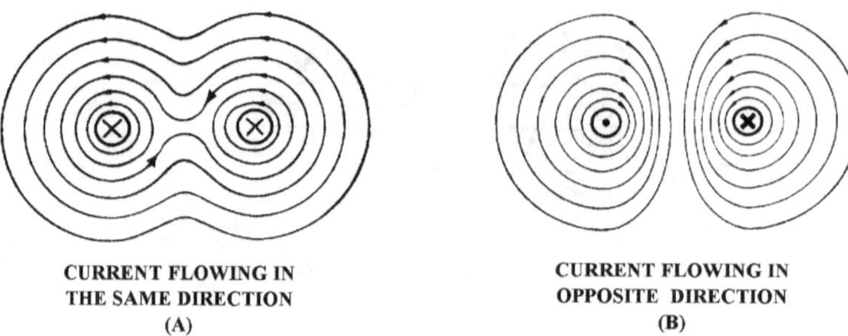

Figure 1-4.—Magnetic field around two parallel conductors.

Q6.   When placed in the vicinity of a current-carrying conductor, the needle of a compass becomes aligned at what angle to the conductor?

Q7.   What is the direction of the magnetic field around a vertical conductor when (a) the current flows upward and (b) the current flows downward.

Q8.   The "left-hand rule" for a conductor is used for what purpose

Q9.   In what direction will the compass needle point when the compass is placed in the magnetic field surrounding a wire?

Q10.   When two adjacent parallel wires carry current in the same direction, the magnetic field about one wire has what effect on the magnetic field about the other conductor?

Q11. When two adjacent parallel conductors carry current in opposite directions, the magnetic field about one conductor has what effect on the magnetic field about the other conductor?

MAGNETIC FIELD OF A COIL

Figure 1-3(A) illustrates that the magnetic field around a current-carrying wire exists at all points along the wire. Figure 1-5 illustrates that when a straight wire is wound around a core, it forms a coil and that the magnetic field about the core assumes a different shape. Figure 1-5(A) is actually a partial cutaway view showing the construction of a simple coil. Figure 1-5(B) shows a cross-sectional view of the same coil. Notice that the two ends of the coil are identified as X and Y.

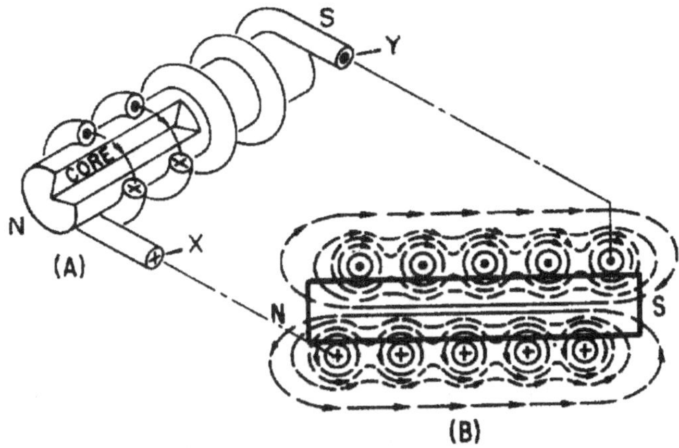

Figure 1-5.—Magnetic field produced by a current-carrying coil.

When current is passed through the coil, the magnetic field about each turn of wire links with the fields of the adjacent turns. (See figure 1-4(A)). The combined influence of all the turns produces a two-pole field similar to that of a simple bar magnet. One end of the coil is a north pole and the other end is a south pole.

Polarity of an Electromagnetic Coil

Figure 1-2 shows that the direction of the magnetic field around a straight wire depends on the direction of current in that wire. Thus, a reversal of current in a wire causes a reversal in the direction of the magnetic field that is produced. It follows that a reversal of the current in a coil also causes a reversal of the two-pole magnetic field about the coil.

When the direction of the current in a coil is known, you can determine the magnetic polarity of the coil by using the LEFT-HAND RULE FOR COILS. This rule, illustrated in figure 1-6, is stated as follows:

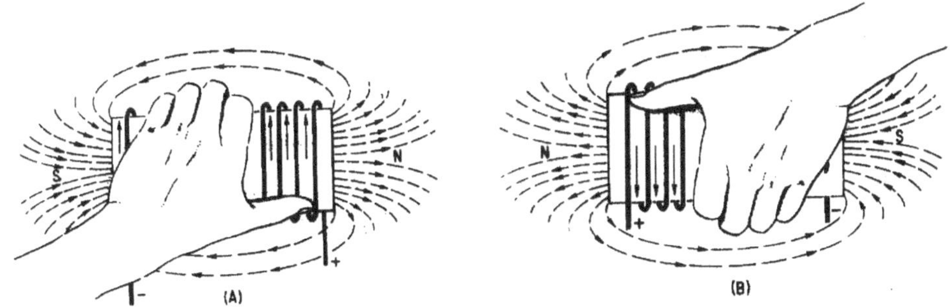

Figure 1-6.—Left-hand rule for coils.

Grasp the coil in your left hand, with your fingers "wrapped around" in the direction of the electron current flow. Your thumb will then point toward the north pole of the coil.

Strength of an Electromagnetic Field

The strength or intensity of a coil's magnetic field depends on a number of factors. The main ones are listed below and will be discussed again later.

- The number of turns of wire in the coil.

- The amount of current flowing in the coil.

- The ratio of the coil length to the coil width.

- The type of material in the core.

Losses in an Electromagnetic Field

When current flows in a conductor, the atoms in the conductor all line up in a definite direction, producing a magnetic field. When the direction of the current changes, the direction of the atoms' alignment also changes, causing the magnetic field to change direction. To reverse all the atoms requires that power be expended, and this power is lost. This loss of power (in the form of heat) is called HYSTERESIS LOSS. Hysteresis loss is common to all ac equipment; however, it causes few problems except in motors, generators, and transformers. When these devices are discussed later in this module, hysteresis loss will be covered in more detail.

Q12. What is the shape of the magnetic field that exists around (a) a straight conductor and (b) a coil?

Q13. What happens to the two-pole field of a coil when the current through the coil is reversed?

Q14. What rule is used to determine the polarity of a coil when the direction of the electron current flow in the coil is known?

Q15. State the rule whose purpose is described in Q14.

BASIC AC GENERATION

From the previous discussion you learned that a current-carrying conductor produces a magnetic field around itself. In module 1, under producing a voltage (emf) using magnetism, you learned how a changing magnetic field produces an emf in a conductor. That is, if a conductor is placed in a magnetic field, and either the field or the conductor moves, an emf is induced in the conductor. This effect is called electromagnetic induction.

CYCLE

Figures 1-7 and 1-8 show a suspended loop of wire (conductor) being rotated (moved) in a clockwise direction through the magnetic field between the poles of a permanent magnet. For ease of explanation, the loop has been divided into a dark half and light half. Notice in (A) of the figure that the dark half is moving along (parallel to) the lines of force. Consequently, it is cutting NO lines of force. The same is true of the light half, which is moving in the opposite direction. Since the conductors are cutting no lines of force, no emf is induced. As the loop rotates toward the position shown in (B), it cuts more and more lines of force per second (inducing an ever-increasing voltage) because it is cutting more directly across the field (lines of force). At (B), the conductor is shown completing one-quarter of a complete revolution, or 90° , of a complete circle. Because the conductor is now cutting directly across the field, the voltage induced in the conductor is maximum. When the value of induced voltage at various points during the rotation from (A) to (B) is plotted on a graph (and the points connected), a curve appears as shown below.

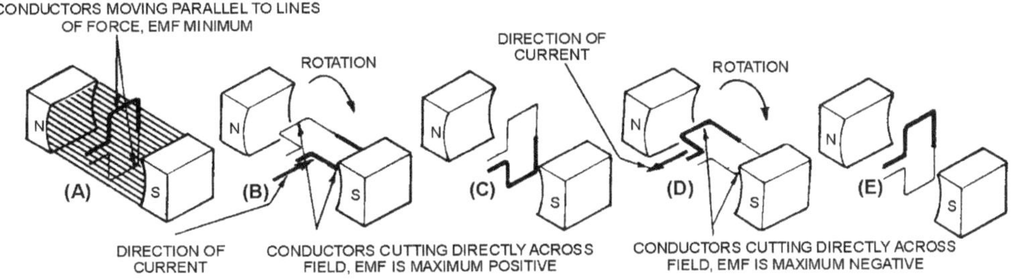

Figure 1-7.—Simple alternating-current generator.

As the loop continues to be rotated toward the position shown below in (C), it cuts fewer and fewer lines of force. The induced voltage decreases from its peak value. Eventually, the loop is once again moving in a plane parallel to the magnetic field, and no emf is induced in the conductor.

The loop has now been rotated through half a circle (one alternation or 180°). If the preceding quarter-cycle is plotted, it appears as shown below.

When the same procedure is applied to the second half of rotation (180° through 360°), the curve appears as shown below. Notice the only difference is in the polarity of the induced voltage. Where previously the polarity was positive, it is now negative.

The sine curve shows the value of induced voltage at each instant of time during rotation of the loop. Notice that this curve contains 360° , or two alternations. TWO ALTERNATIONS represent ONE complete CYCLE of rotation.

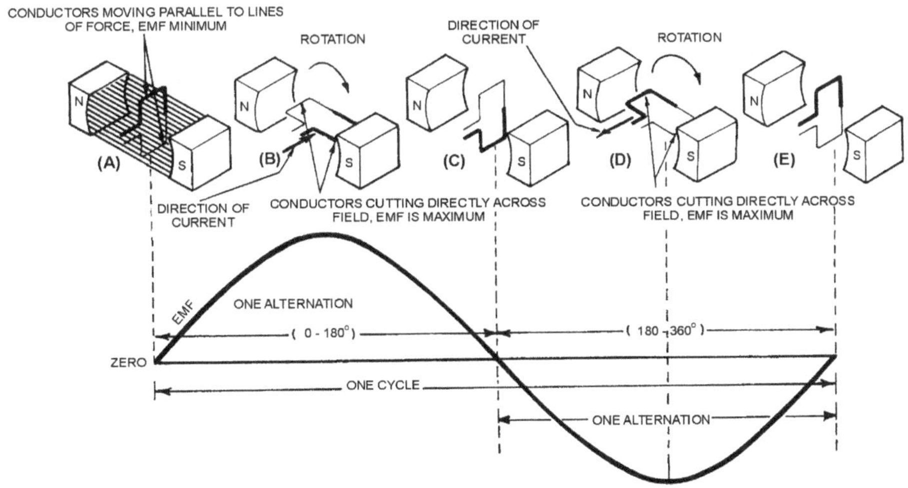

Figure 1-8.—Basic alternating-current generator.

Assuming a closed path is provided across the ends of the conductor loop, you can determine the direction of current in the loop by using the LEFT-HAND RULE FOR GENERATORS. Refer to figure 1-9. The left-hand rule is applied as follows: First, place your left hand on the illustration with the fingers as shown. Your THUMB will now point in the direction of rotation (relative movement of the wire to the magnetic field); your FOREFINGER will point in the direction of magnetic flux (north to south); and your MIDDLE FINGER (pointing out of the paper) will point in the direction of electron current flow.

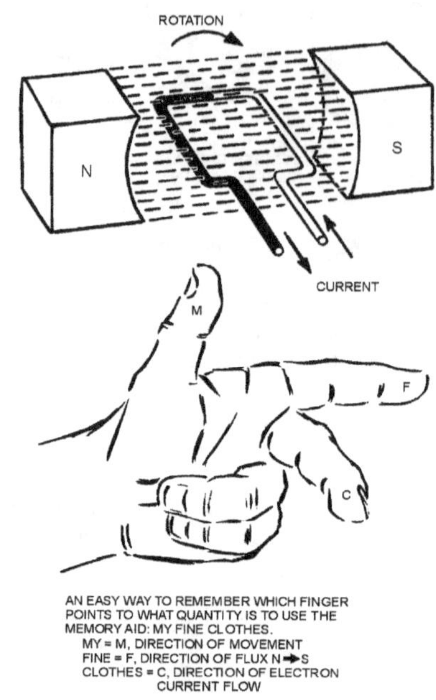

Figure 1-9.—Left-hand rule for generators.

By applying the left-hand rule to the dark half of the loop in (B) in figure 1-8, you will find that the current flows in the direction indicated by the heavy arrow. Similarly, by using the left-hand rule on the light half of the loop, you will find that current therein flows in the opposite direction. The two induced voltages in the loop add together to form one total emf. It is this emf which causes the current in the loop.

When the loop rotates to the position shown in (D) of figure 1-8, the action reverses. The dark half is moving up instead of down, and the light half is moving down instead of up. By applying the left-hand rule once again, you will see that the total induced emf and its resulting current have reversed direction. The voltage builds up to maximum in this new direction, as shown by the sine curve in figure 1-8. The loop finally returns to its original position (E), at which point voltage is again zero. The sine curve represents one complete cycle of voltage generated by the rotating loop. All the illustrations used in this chapter show the wire loop moving in a clockwise direction. In actual practice, the loop can be moved clockwise or counterclockwise. Regardless of the direction of movement, the left-hand rule applies.

If the loop is rotated through 360° at a steady rate, and if the strength of the magnetic field is uniform, the voltage produced is a sine wave of voltage, as indicated in figure 1-9. Continuous rotation of the loop will produce a series of sine-wave voltage cycles or, in other words, an ac voltage.

As mentioned previously, the cycle consists of two complete alternations in a period of time. Recently the HERTZ (Hz) has been designated to indicate one cycle per second. If ONE CYCLE PER SECOND is ONE HERTZ, then 100 cycles per second are equal to 100 hertz, and so on. Throughout the NEETS, the term cycle is used when no specific time element is involved, and the term hertz (Hz) is used when the time element is measured in seconds.

Q16.   When a conductor is rotated in a magnetic field, at what points in the cycle is emf (a) at maximum amplitude and (b) at minimum amplitude?

Q17.   One cycle is equal to how many degrees of rotation of a conductor in a magnetic field?

Q18.   State the left-hand rule used to determine the direction of current in a generator.

Q19.   How is an ac voltage produced by an ac generator?

FREQUENCY

If the loop in the figure 1-8 (A) makes one complete revolution each second, the generator produces one complete cycle of ac during each second (1 Hz). Increasing the number of revolutions to two per second will produce two complete cycles of ac per second (2 Hz). The number of complete cycles of alternating current or voltage completed each second is referred to as the FREQUENCY. Frequency is always measured and expressed in hertz.

Alternating-current frequency is an important term to understand since most ac electrical equipments require a specific frequency for proper operation.

Q20.   Define Frequency.

PERIOD

An individual cycle of any sine wave represents a definite amount of TIME. Notice that figure 1-10 shows 2 cycles of a sine wave which has a frequency of 2 hertz (Hz). Since 2 cycles occur each second, 1 cycle must require one-half second of time. The time required to complete one cycle of a waveform is called the PERIOD of the wave. In figure 1-10, the period is one-half second. The relationship between time (t) and frequency (f) is indicated by the formulas

$$t = \frac{1}{f} \quad \text{and} \quad f = \frac{1}{t}$$

where t = period in seconds and
f = frequency in hertz

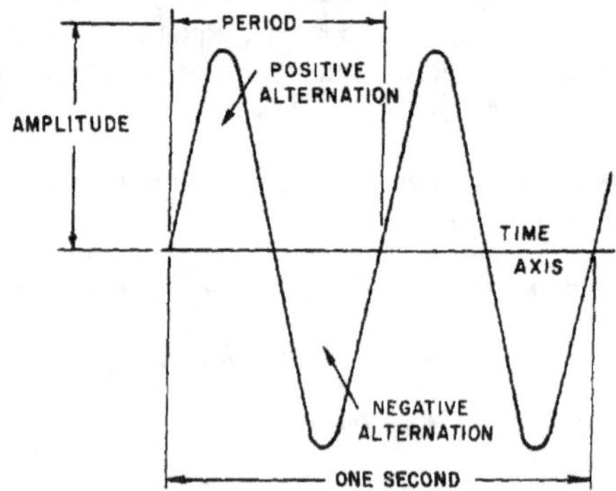

Figure 1-10.—Period of a sine wave.

Each cycle of the sine wave shown in figure 1-10 consists of two identically shaped variations in voltage. The variation which occurs during the time the voltage is positive is called the POSITIVE ALTERNATION. The variation which occurs during the time the voltage is negative is called the NEGATIVE ALTERNATION. In a sine wave, these two alternations are identical in size and shape, but opposite in polarity.

The distance from zero to the maximum value of each alternation is called the AMPLITUDE. The amplitude of the positive alternation and the amplitude of the negative alternation are the same.

WAVELENGTH

The time it takes for a sine wave to complete one cycle is defined as the period of the waveform. The distance traveled by the sine wave during this period is referred to as WAVELENGTH. Wavelength, indicated by the symbol λ (Greek lambda), is the distance along the waveform from one point to the same point on the next cycle. You can observe this relationship by examining figure 1-11. The point on the waveform that measurement of wavelength begins is not important as long as the distance is measured to the same point on the next cycle (see figure 1-12).

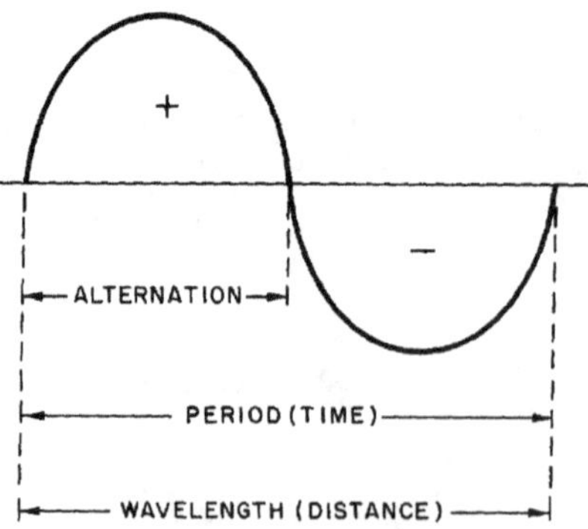

Figure 1-11.—Wavelength.

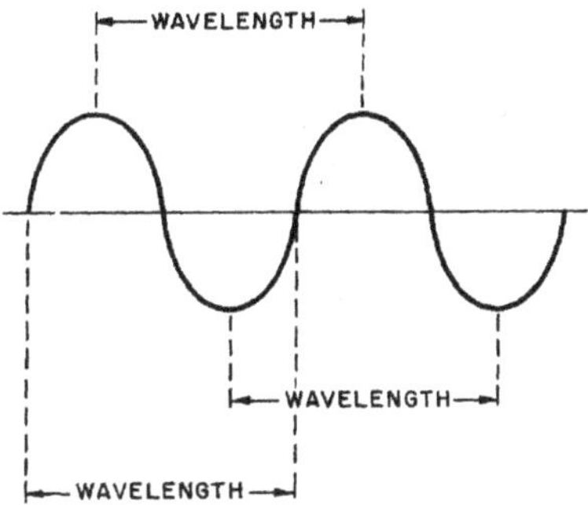

Figure 1-12.—Wavelength measurement.

Q21. What term is used to indicate the time of one complete cycle of a waveform?

Q22. What is a positive alternation?

Q23. What do the period and the wavelength of a sine wave measure, respectively?

ALTERNATING CURRENT VALUES

In discussing alternating current and voltage, you will often find it necessary to express the current and voltage in terms of MAXIMUM or PEAK values, PEAK-to-PEAK values, EFFECTIVE values, AVERAGE values, or INSTANTANEOUS values. Each of these values has a different meaning and is used to describe a different amount of current or voltage.

PEAK AND PEAK-TO-PEAK VALUES

Refer to figure 1-13. Notice it shows the positive alternation of a sine wave (a half-cycle of ac) and a dc waveform that occur simultaneously. Note that the dc starts and stops at the same moment as does the positive alternation, and that both waveforms rise to the same maximum value. However, the dc values are greater than the corresponding ac values at all points except the point at which the positive alternation passes through its maximum value. At this point the dc and ac values are equal. This point on the sine wave is referred to as the maximum or peak value.

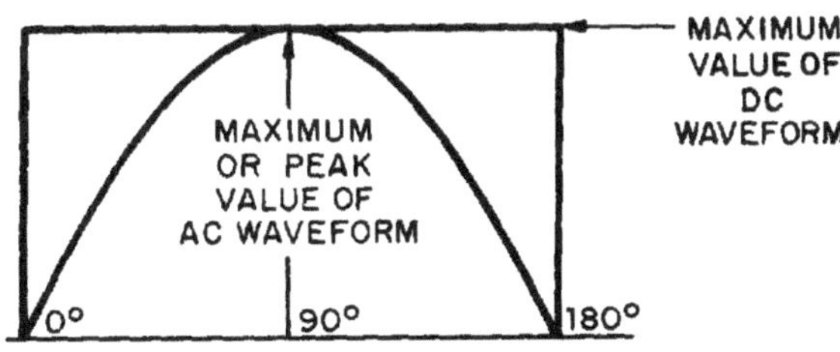

Figure 1-13.—Maximum or peak value.

During each complete cycle of ac there are always two maximum or peak values, one for the positive half-cycle and the other for the negative half-cycle. The difference between the peak positive value and the peak negative value is called the peak-to-peak value of the sine wave. This value is twice the maximum or peak value of the sine wave and is sometimes used for measurement of ac voltages. Note the difference between peak and peak-to-peak values in figure 1-14. Usually alternating voltage and current are expressed in EFFECTIVE VALUES (a term you will study later) rather than in peak-to-peak values.

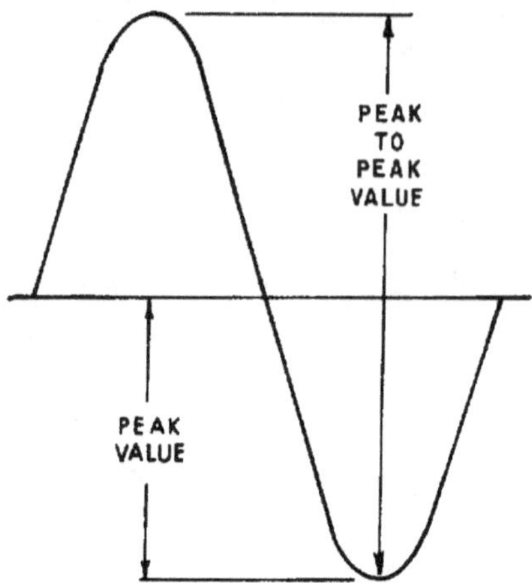

Figure 1-14.—Peak and peak-to-peak values.

Q24. What is meant by peak and peak-to-peak values of ac?

Q25. How many times is the maximum or peak value of emf or current reached during one cycle of ac?

INSTANTANEOUS VALUE

The INSTANTANEOUS value of an alternating voltage or current is the value of voltage or current at one particular instant. The value may be zero if the particular instant is the time in the cycle at which the polarity of the voltage is changing. It may also be the same as the peak value, if the selected instant is the time in the cycle at which the voltage or current stops increasing and starts decreasing. There are actually an infinite number of instantaneous values between zero and the peak value.

AVERAGE VALUE

The AVERAGE value of an alternating current or voltage is the average of ALL the INSTANTANEOUS values during ONE alternation. Since the voltage increases from zero to peak value and decreases back to zero during one alternation, the average value must be some value between those two limits. You could determine the average value by adding together a series of instantaneous values of the alternation (between 0° and 180°), and then dividing the sum by the number of instantaneous values used. The computation would show that one alternation of a sine wave has an average value equal to 0.636 times the peak value. The formula for average voltage is

$$E_{avg} = 0.636 \times E_{max}$$

where E_{avg} is the average voltage of one alternation, and E_{max} is the maximum or peak voltage. Similarly, the formula for average current is

$$I_{avg} = 0.636 \times I_{max}$$

where I_{avg} is the average current in one alternation, and I_{max} is the maximum or peak current.

Do not confuse the above definition of an average value with that of the average value of a complete cycle. Because the voltage is positive during one alternation and negative during the other alternation, the average value of the voltage values occurring during the complete cycle is <u>zero</u>.

Q26. *If any point on a sine wave is selected at random and the value of the current or voltage is measured at that one particular moment, what value is being measured?*

Q27. *What value of current or voltage is computed by averaging all of the instantaneous values during the negative alternation of a sine wave?*

Q28. *What is the average value of all of the instantaneous currents or voltages occurring during one complete cycle of a sine wave?*

Q29. *What mathematical formulas are used to find the average value of current and average value of voltage of a sine wave?*

Q30. *If E_{max} is 115 volts, what is E_{avg}?*

Q31. *If I_{avg} is 1.272 ampere, what is I_{max}?*

EFFECTIVE VALUE OF A SINE WAVE

E_{max}, E_{avg}, I_{max}, and I_{avg} are values used in ac measurements. Another value used is the EFFECTIVE value of ac This is the value of alternating voltage or current that will have the same effect on a resistance as a comparable value of direct voltage or current will have on the same resistance.

In an earlier discussion you were told that when current flows in a resistance, heat is produced. When direct current flows in a resistance, the amount of electrical power converted into heat equals I^2R watts. However, since an alternating current having a maximum value of 1 ampere does not maintain a constant value, the alternating current will not produce as much heat in the resistance as will a direct current of 1 ampere.

Figure 1-15 compares the heating effect of 1 ampere of dc to the heating effect of 1 ampere of ac.

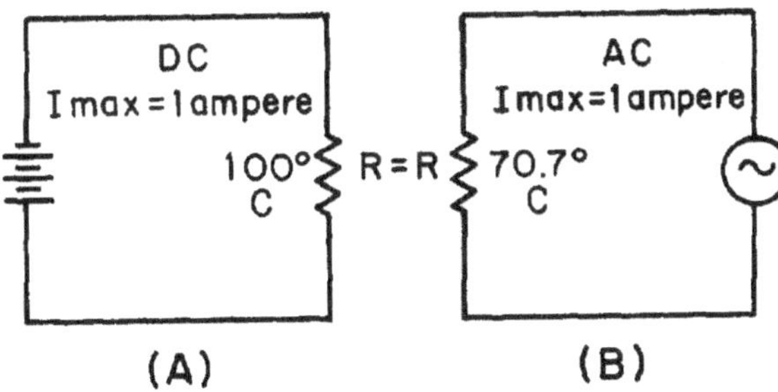

Figure 1-15.—Heating effect of ac and dc.

Examine views A and B of figure 1-15 and notice that the heat (70.7° C) produced by 1 ampere of alternating current (that is, an ac with a maximum value of 1 ampere) is only 70.7 percent of the heat (100° C) produced by 1 ampere of direct current. Mathematically,

$$\frac{\text{The heating effect of 1 maximum a.c. ampere}}{\text{The heating effect of 1 maximum d.c. ampere}} = \frac{70.7°\,C}{100°\,C} = 0.707$$

Therefore, for effective value of ac $(I_{eff}) = 0.707 \times I_{max}$.

The rate at which heat is produced in a resistance forms a convenient basis for establishing an effective value of alternating current, and is known as the "heating effect" method. An alternating current is said to have an effective value of one ampere when it produces heat in a given resistance at the same rate as does one ampere of direct current.

You can compute the effective value of a sine wave of current to a fair degree of accuracy by taking equally-spaced instantaneous values of current along the curve and extracting the square root of the average of the sum of the squared values.

For this reason, the effective value is often called the "root-mean-square" (rms) value. Thus,

$$I_{eff} = \sqrt{\text{Average of the sum of the squares of } I_{inst}}.$$

Stated another way, the effective or rms value (I_{eff}) of a sine wave of current is 0.707 times the maximum value of current (I_{max}). Thus, $I_{eff} = 0.707 \times I_{max}$. When I_{eff} is known, you can find I_{max} by using the formula $I_{max} = 1.414 \times I_{eff}$. You might wonder where the constant 1.414 comes from. To find out, examine figure 1-15 again and read the following explanation. Assume that the dc in figure 1-15(A) is maintained at 1 ampere and the resistor temperature at 100° C. Also assume that the ac in figure 1-15(B) is increased until the temperature of the resistor is 100° C. At this point it is found that a maximum ac value of 1.414 amperes is required in order to have the same heating effect as direct current. Therefore, in the ac circuit the maximum current required is 1.414 times the effective current. It is important for you to remember the above relationship and that the effective value (I_{eff}) of any sine wave of current is always 0.707 times the maximum value (I_{max}).

Since alternating current is caused by an alternating voltage, the ratio of the effective value of voltage to the maximum value of voltage is the same as the ratio of the effective value of current to the maximum value of current. Stated another way, the effective or rms value (E_{eff}) of a sine-wave of voltage is 0.707 times the maximum value of voltage (E_{max}),

Thus,

$$E_{eff} = \sqrt{\text{Average of the sum of the squares of } E_{inst}}$$

or,

$$E_{eff} = 0.707 \times E_{max}$$

and,

$$E_{max} = 1.414 \times E_{eff}$$

When an alternating current or voltage value is specified in a book or on a diagram, the value is an effective value unless there is a definite statement to the contrary. Remember that all meters, unless marked to the contrary, are calibrated to indicate effective values of current and voltage.

Problem: A circuit is known to have an alternating voltage of 120 volts and a peak or maximum current of 30 amperes. What are the peak voltage and effective current values?

Given:
$$E_s = 120 \text{ V}$$
$$E_{max} = 30 \text{ A}$$

Solution:
$$E_{max} = 1.414 \times E_{eff}$$
$$E_{max} = 1.414 \times 120 \text{ volts}$$
$$E_{max} = 169.68 \text{ volts}$$
$$I_{eff} = 0.707 \times I_{max}$$
$$I_{eff} = 0.707 \times 30 \text{ amperes}$$
$$I_{eff} = 21.21 \text{ amperes}$$

Figure 1-16 shows the relationship between the various values used to indicate sine-wave amplitude. Review the values in the figure to ensure you understand what each value indicates.

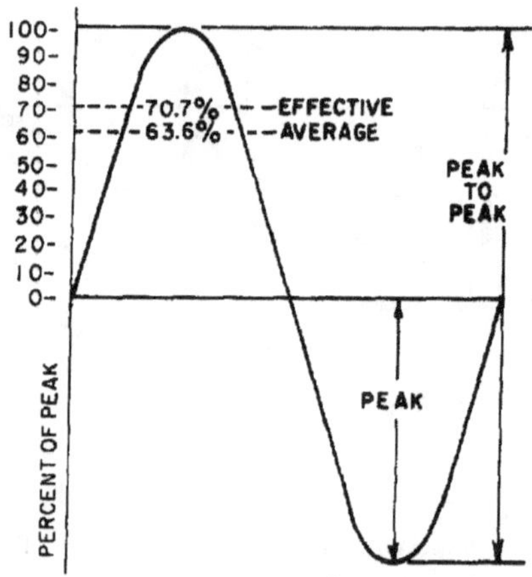

Figure 1-16.—Various values used to indicate sine-wave amplitude.

Q32. What is the most convenient basis for comparing alternating and direct voltages and currents?

Q33. What value of ac is used as a comparison to dc?

Q34. What is the formula for finding the effective value of an alternating current?

Q35. If the peak value of a sine wave is 1,000 volts, what is the effective (E_{eff}) value?

Q36. If I_{eff} = 4.25 ampere, what is I_{max}?

SINE WAVES IN PHASE

When a sine wave of voltage is applied to a resistance, the resulting current is also a sine wave. This follows Ohm's law which states that current is directly proportional to the applied voltage. Now examine figure 1-17. Notice that the sine wave of voltage and the resulting sine wave of current are superimposed on the same time axis. Notice also that as the voltage increases in a positive direction, the current increases along with it, and that when the voltage reverses direction, the current also reverses direction. When two sine waves, such as those represented by figure 1-17, are precisely in step with one another, they are said to be IN PHASE. To be in phase, the two sine waves must go through their maximum and minimum points at the same time and in the same direction.

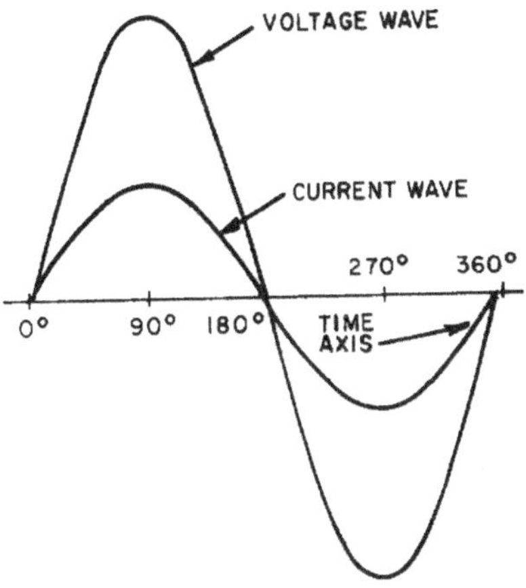

Figure 1-17.—Voltage and current waves in phase.

In some circuits, several sine waves can be in phase with each other. Thus, it is possible to have two or more voltage drops in phase with each other and also be in phase with the circuit current.

SINE WAVES OUT OF PHASE

Figure 1-18 shows voltage wave E_1 which is considered to start at 0° (time one). As voltage wave E_1 reaches its positive peak, voltage wave E_2 starts its rise (time two). Since these voltage waves do not go through their maximum and minimum points at the same instant of time, a PHASE DIFFERENCE exists between the two waves. The two waves are said to be OUT OF PHASE. For the two waves in figure 1-18 the phase difference is 90° .

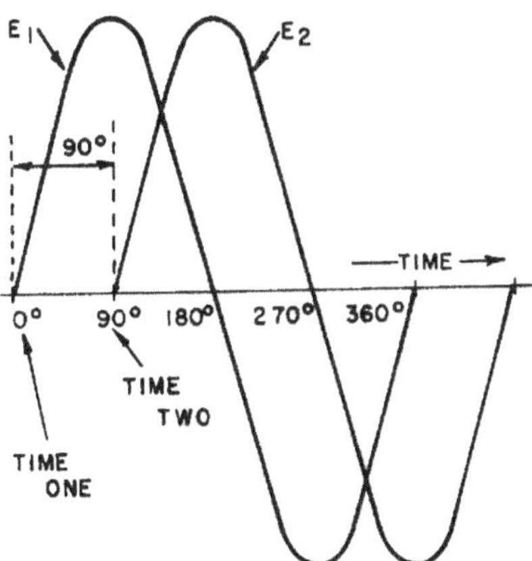

Figure 1-18.—Voltage waves 90° out of phase.

To further describe the phase relationship between two sine waves, the terms LEAD and LAG are used. The amount by which one sine wave leads or lags another sine wave is measured in degrees. Refer again to figure 1-18. Observe that wave E_2 starts 90° later in time than does wave E_1. You can also describe this relationship by saying that wave E_1 leads wave E_2 by 90°, or that wave E_2 lags wave E_1 by 90°. (Either statement is correct; it is the phase relationship between the two sine waves that is important.)

It is possible for one sine wave to lead or lag another sine wave by any number of degrees, except 0° or 360°. When the latter condition exists, the two waves are said to be in phase. Thus, two sine waves that differ in phase by 45° are actually out of phase with each other, whereas two sine waves that differ in phase by 360° are considered to be in phase with each other.

A phase relationship that is quite common is shown in figure 1-19. Notice that the two waves illustrated differ in phase by 180°. Notice also that although the waves pass through their maximum and minimum values at the same time, their instantaneous voltages are always of opposite polarity. If two such waves exist across the same component, and the waves are of equal amplitude, they cancel each other. When they have different amplitudes, the resultant wave has the same polarity as the larger wave and has an amplitude equal to the difference between the amplitudes of the two waves.

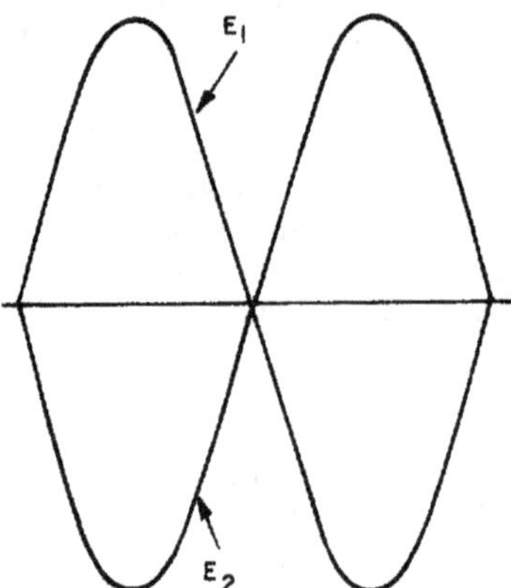

Figure 1-19.—Voltage waves 180° out of phase.

To determine the phase difference between two sine waves, locate the points on the time axis where the two waves cross the time axis traveling in the same direction. The number of degrees between the crossing points is the phase difference. The wave that crosses the axis at the later time (to the right on the time axis) is said to lag the other wave.

Q37. When are the voltage wave and the current wave in a circuit considered to be in phase?

Q38. When are two voltage waves considered to be out of phase?

Q39. What is the phase relationship between two voltage waves that differ in phase by 360°?

Q40. How do you determine the phase difference between two sine waves that are plotted on the same graph?

OHM'S LAW IN AC CIRCUITS

Many ac circuits contain resistance only. The rules for these circuits are the same rules that apply to dc circuits. Resistors, lamps, and heating elements are examples of resistive elements. When an ac circuit contains only resistance, Ohm's Law, Kirchhoff's Law, and the various rules that apply to voltage, current, and power in a dc circuit also apply to the ac circuit. The Ohm's Law formula for an ac circuit can be stated as

$$I_{eff} = \frac{E_{eff}}{R} \text{ or } I = \frac{E}{R}$$

Remember, unless otherwise stated, all ac voltage and current values are given as effective values. The formula for Ohm's Law can also be stated as

$$I_{avg} = \frac{E_{avg}}{R} \text{ or } I_{max} = \frac{E_{max}}{R}$$

$$I_{peak-to-peak} = \frac{E_{peak-to-peak}}{R}$$

The important thing to keep in mind is: <u>Do Not mix ac values</u>. When you solve for effective values, all values you use in the formula <u>must be effective values</u>. Similarly, when you solve for average values, all values you use <u>must be average values</u>. This point should be clearer after you work the following problem: A series circuit consists of two resistors (R1 = 5 ohms and R2 = 15 ohms) and an alternating voltage source of 120 volts. What is I_{avg}?

Given: R1 = 5 ohms

               R2 = 15 ohms

               E_s = 120 ohms

Solution: First solve for total resistance R_T.

$$R_T = R1 + R2$$
$$R_T = 5 \text{ ohms} + 15 \text{ ohms}$$
$$R_T = 20 \text{ ohms}$$

The alternating voltage is assumed to be an effective value (since it is not specified to be otherwise). Apply the Ohm's Law formula.

$$I_{eff} = \frac{E_{eff}}{R}$$

$$I_{eff} = \frac{120 \text{ volts}}{20 \text{ ohms}}$$

$$I_{eff} = 6 \text{ amperes}$$

The problem, however, asked for the average value of current (I_{avg}). To convert the effective value of current to the average value of current, you must first determine the peak or maximum value of current, I_{max}.

$$I_{max} = 1.414 \times I_{eff}$$

$$I_{max} = 1.414 \times 6 \text{ amperes}$$

$$I_{max} = 8.484 \text{ amperes}$$

You can now find I_{avg}. Just substitute 8.484 amperes in the I_{avg} formula and solve for I_{avg}.

$$I_{avg} = 0.636 \times I_{max}$$

$$I_{avg} = 0.636 \times 8.484 \text{ amperes}$$

$$I_{avg} = 5.4 \text{ amperes (rounded off to one decimal place)}$$

Remember, you can use the Ohm's Law formulas to solve any <u>purely resistive</u> ac circuit problem. Use the formulas in the same manner as you would to solve a dc circuit problem.

Q41. *A series circuit consists of three resistors (R1 = 10Ω, R2 = 20Ω, R3 = 15Ω) and an alternating voltage source of 100 volts. What is the effective value of current in the circuit?*

Q42. *If the alternating source in Q41 is changed to 200 volts peak-to-peak, what is I_{avg}?*

Q43. *If E_{eff} is 130 volts and I_{eff} is 3 amperes, what is the total resistance (R_T) in the circuit?*

SUMMARY

Before going on to chapter 2, read the following summary of the material in chapter 1. This summary will reinforce what you have already learned.

<u>DC AND AC</u>—Direct current flows in one direction only, while alternating current is constantly changing in amplitude and direction.

<u>ADVANTAGES AND DISADVANTAGES OF AC AND DC</u>—Direct current has several disadvantages compared to alternating current. Direct current, for example, must be generated at the voltage level required by the load. Alternating current, however, can be generated at a high level and stepped down at the consumer end (through the use of a transformer) to whatever voltage level is required by the load. Since power in a dc system must be transmitted at low voltage and high current levels, the I^2R power loss becomes a problem in the dc system. Since power in an ac system can be transmitted at a high voltage level and a low current level, the I^2R power loss in the ac system is much less than that in the dc system.

VOLTAGE WAVEFORMS—The waveform of voltage or current is a graphical picture of changes in voltage or current values over a period of time.

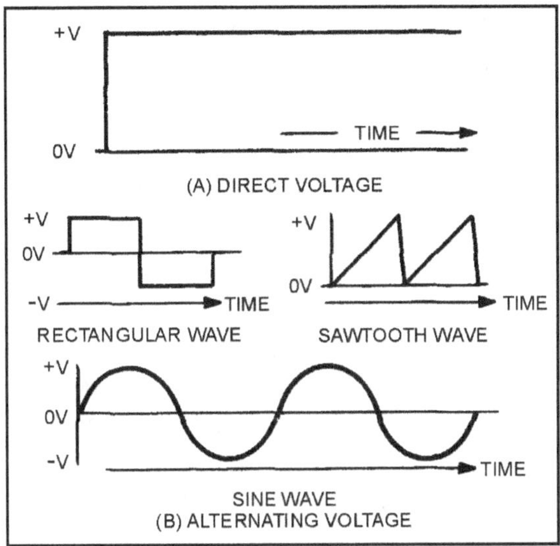

ELECTROMAGNETISM—When a compass is placed in the vicinity of a current-carrying conductor, the needle aligns itself at right angles to the conductor. The north pole of the compass indicates the direction of the magnetic field produced by the current. <u>By knowing</u> the direction of current, you can use the left-hand rule for conductors to determine the direction of the magnetic lines of force.

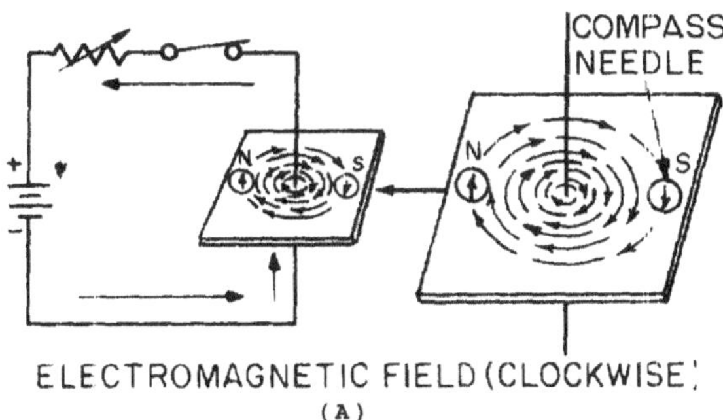

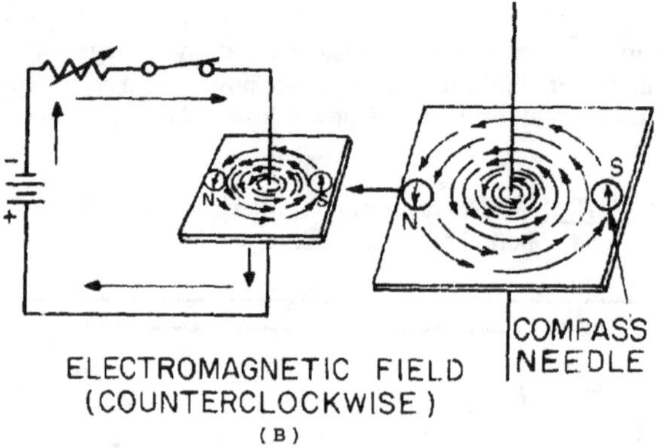

Arrows are generally used in electrical diagrams to indicate the direction of current in a wire. A cross (+) on the end of a cross-sectional view of a wire indicates that current is flowing away from you, while a dot (·) indicates that current is flowing toward you.

When two adjacent parallel conductors carry current in the same direction, the magnetic fields around the conductors aid each other. When the currents in the two conductors flow in opposite directions, the fields around the conductors oppose each other.

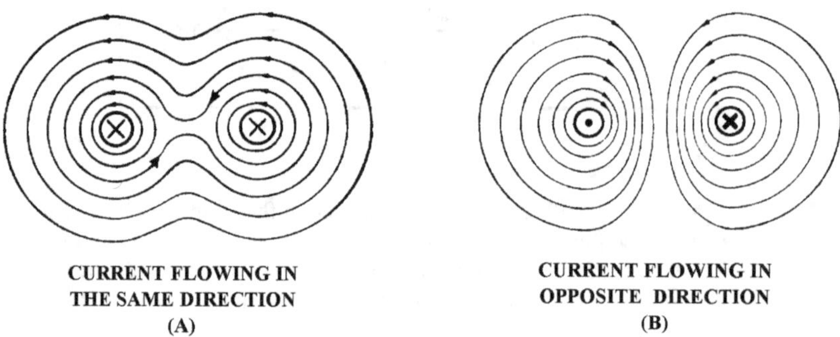

**CURRENT FLOWING IN
THE SAME DIRECTION
(A)**

**CURRENT FLOWING IN
OPPOSITE DIRECTION
(B)**

MAGNETIC FIELD OF A COIL—When wire is wound around a core, it forms a COIL. The magnetic fields produced when current flows in the coil combine. The combined influence of all of the fields around the turns produce a two-pole field similar to that of a simple bar magnet.

When the direction of current in the coil is reversed, the polarity of the two-pole field of the coil is reversed.

The strength of the magnetic field of the coil is dependent upon:

- The number of turns of the wire in the coil.

- The amount of current in the coil.

- The ratio of the coil length to the coil width.

- The type of material in the core.

BASIC AC GENERATION—When a conductor is in a magnetic field and either the field or the conductor moves, an emf (voltage) is induced in the conductor. This effect is called electromagnetic induction.

A loop of wire rotating in a magnetic field produces a voltage which constantly changes in amplitude and direction. The waveform produced is called a sine wave and is a graphical picture of alternating current (ac). One complete revolution (360°) of the conductor produces one cycle of ac. The cycle is composed of two alternations: a positive alternation and a negative alternation. One cycle of ac in one second is equal to 1 hertz (1 Hz).

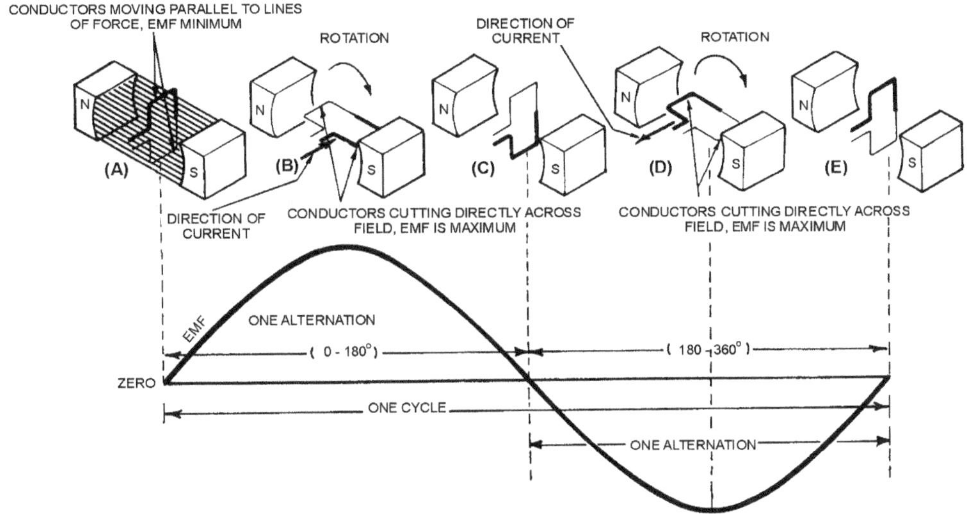

FREQUENCY—The number of cycles of ac per second is referred to as the FREQUENCY. AC frequency is measured in hertz. Most ac equipment is rated by frequency as well as by voltage and current.

PERIOD—The time required to complete one cycle of a waveform is called the PERIOD OF THE WAVE.

Each ac sine wave is composed of two alternations. The alternation which occurs during the time the sine wave is positive is called the positive alternation. The alternation which occurs during the time the sine wave is negative is called the negative alternation. In each cycle of sine wave, the two alternations are identical in size and shape, but opposite in polarity.

The period of a sine wave is inversely proportional to the frequency; e.g., the higher the frequency, the shorter the period. The mathematical relationships between time and frequency are

$$t = \frac{1}{f} \text{ and } f = \frac{1}{t}$$

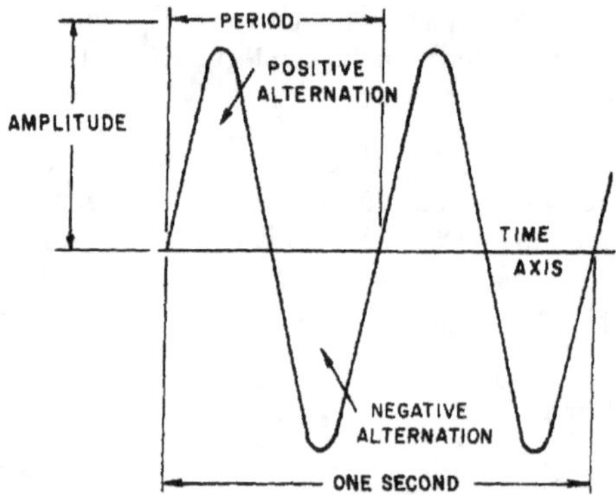

WAVELENGTH—The period of a sine wave is defined as the time it takes to complete one cycle. The distance the waveform covers during this period is referred to as the wavelength. Wavelength is indicated by lambda (λ) and is measured from a point on a given waveform (sine wave) to the corresponding point on the next waveform.

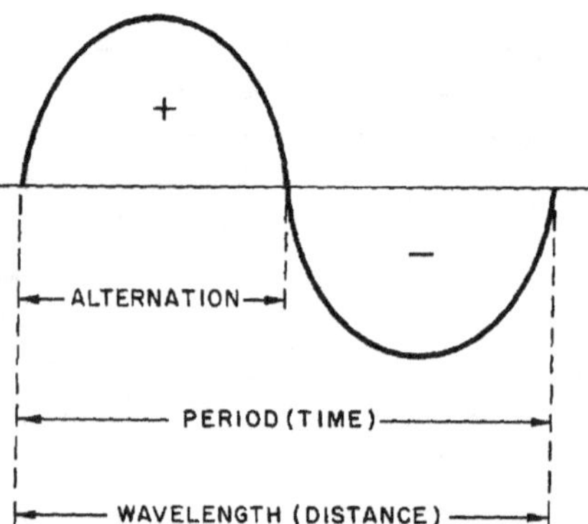

PEAK AND PEAK-TO-PEAK VALUES—The maximum value reached during one alternation of a sine wave is the peak value. The maximum reached during the positive alternation to the maximum value reached during the negative alternation is the peak-to-peak value. The peak-to-peak value is twice the peak value.

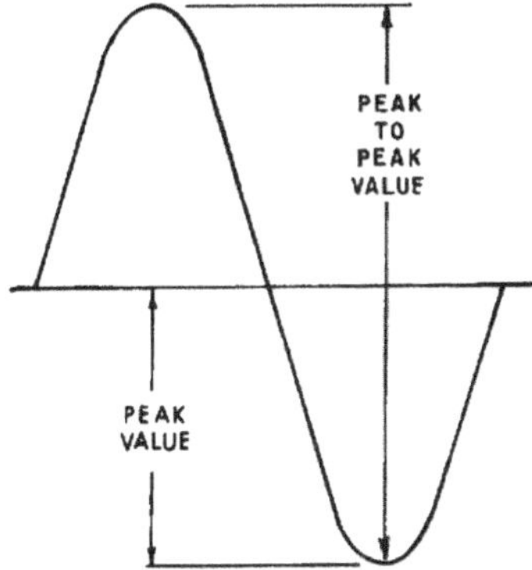

INSTANTANEOUS VALUE—The instantaneous value of a sine wave of alternating voltage or current is the value of voltage or current at one particular instant of time. There are an infinite number of instantaneous values between zero and the peak value.

AVERAGE VALUE—The average value of a sine wave of voltage or current is the average of all the instantaneous values during one alternation. The average value is equal to 0.636 of the peak value. The formulas for average voltage and average current are:

$$E_{avg} = 0.636 \times E_{max}$$
$$I_{avg} = 0.636 \times I_{max}$$

Remember: The average value (E_{avg} or I_{avg}) is for one alternation only. The average value of a complete sine wave is zero.

EFFECTIVE VALUE—The effective value of an alternating current or voltage is the value of alternating current or voltage that produces the same amount of heat in a resistive component that would be produced in the same component by a direct current or voltage of the same value. The effective value of a sine wave is equal to 0.707 times the peak value. The effective value is also called the root mean square or rms value.

The term rms value is used to describe the process of determining the effective value of a sine wave by using the instantaneous value of voltage or current. You can find the rms value of a current or voltage by taking equally spaced instantaneous values on the sine wave and extracting the square root of the average of the sum of the instantaneous values. This is where the term "Root-Mean-Square" (rms) value comes from.

The formulas for effective and maximum values of voltage and current are:

$$E_{eff} = 0.707 \times E_{max}$$
$$E_{max} = 1.414 \times E_{eff}$$
$$I_{eff} = 0.707 \times I_{max}$$
$$I_{max} = 1.414 \times I_{eff}$$

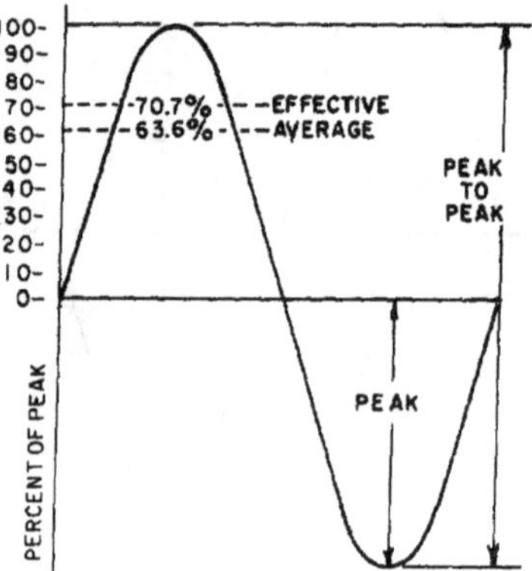

SINE WAVES IN PHASE—When two sine waves are exactly in step with each other, they are said to be in phase. To be in phase, both sine waves must go through their minimum and maximum points at the same time and in the same direction.

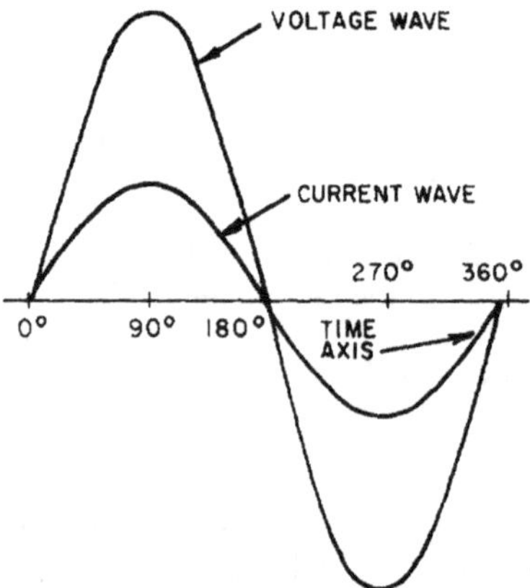

SINE WAVES OUT OF PHASE—When two sine waves go through their minimum and maximum points at different times, a phase difference exists between them. The two waves are said to be out of phase with each other. To describe this phase difference, the terms lead and lag are used. The wave that reaches its minimum (or maximum) value first is said to lead the other wave. The term lag is used to describe the wave that reaches its minimum (or maximum) value some time after the first wave does. When a sine wave is described as leading or lagging, the difference in degrees is usually stated. For example, wave E_1 leads wave E_2 by 90°, or wave E_2 lags wave E_1 by 90°. Remember: Two sine waves can differ by any number of degrees except 0° and 360°. Two sine waves that differ by 0° or by 360° are considered to be in phase. Two sine waves that are opposite in polarity and that differ by 180° are said to be out of phase, even though they go through their minimum and maximum points at the same time.

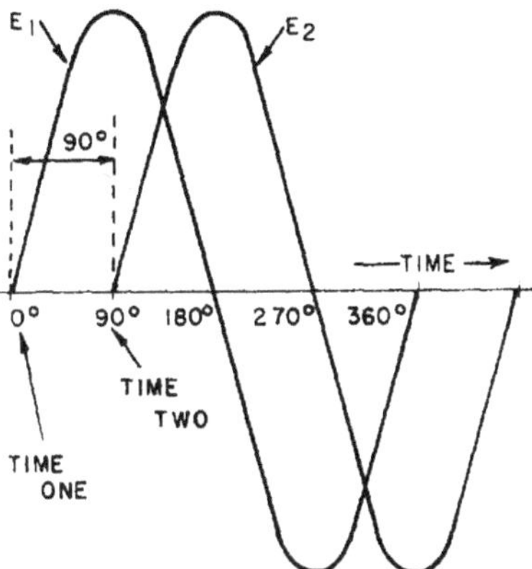

OHM'S LAW IN AC CIRCUIT—All dc rules and laws apply to an ac circuit that contains <u>only</u> <u>resistance</u>. The important point to remember is: Do not <u>mix ac values</u>. Ohm's Law formulas for ac circuits are given below:

$$I = \frac{E}{R}$$

$$I_{eff} = \frac{E_{eff}}{R}$$

$$I_{avg} = \frac{E_{avg}}{R}$$

$$I_{max} = \frac{E_{max}}{R}$$

$$I_{Peak-to-Peak} = \frac{E_{Peak-to-Peak}}{R}$$

ANSWERS TO QUESTIONS Q1. THROUGH Q43.

A1.   *An electrical current which flows in one direction only.*

A2.   *An electrical current which is constantly varying in amplitude, and which changes direction at regular intervals.*

A3.   *The dc voltage must be generated at the level required by the load.*

A4.   *The I^2R power loss is excessive.*

A5.   *Alternating current (ac).*

A6.   *The needle aligns itself at right angles to the conductor.*

A7.   *(a) clockwise (b) counterclockwise.*

A8.   *It is used to determine the relation between the direction of the magnetic lines of force around a conductor and the direction of current through the conductor.*

A9.   *The north pole of the compass will point in the direction of the magnetic lines of force.*

A10.   *It combines with the other field.*

A11.   *It deforms the other field.*

A12.   *(a) The field consists of concentric circles in a plane perpendicular to the wire (b) the field of each turn of wire links with the fields of adjacent turns producing a two-pole field similar in shape to that of a simple bar magnet.*

A13.   *The polarity of the two-pole field reverses.*

A14.   *Use the left-hand rule for coils.*

A15.   *Grasp the coil in your left hand, with your fingers "wrapped around" in the direction of electron flow. The thumb will point toward the north pole.*

A16.   *(a) When the conductors are cutting directly across the magnetic lines of force (at the 90° and 270° points). (b) When the conductors are moving parallel to the magnetic lines of force (at the 0°, 180°, and 360° points).*

A17.   *360°.*

A18.   *Extend your left hand so that your thumb points in the direction of conductor movement, and your forefinger points in the direction of the magnetic flux (north to south). Now point your middle finger 90° from the forefinger and it will point in the direction of electron current flow in the conductor.*

A19.   *Continuous rotation of the conductor through magnetic fines of force produces a series of cycles of alternating voltage or, in other words, an alternating voltage or a sine wave of voltage.*

A20.   *Frequency is the number of complete cycles of alternating voltage or current completed each second.*

A21.   *Period.*

A22.   *A positive alternation is the positive variation in the voltage or current of a sine curve.*

A23.   *The period measures time and the wavelength measures distance.*

A24.   *The peak value is the maximum value of one alternation; the peak-to-peak value is twice the maximum or peak value.*

A25.   *Twice.*

A26.   *The instantaneous value (E_{inst} or I_{inst})*

A27.   *Average value (E_{avg} or I_{avg})*

A28.   *Zero*

A29.

$$I_{avg} = 0.636 \times I_{max}$$

$$E_{avg} = 0.636 \times E_{max}$$

A30.

$$E_{avg} = 0.636 \times 115 \text{ volts}$$

$$E_{avg} = 73.14 \text{ volts}$$

A31.

$$\text{If } I_{avg} = I_{max} \times 0.636, \text{ then } I_{max} = \frac{I_{avg}}{0.636}$$

Thus,

$$I_{max} = \frac{1.272}{0.636} \text{ ampere} = 2 \text{ amperes}$$

A32.   *The power (heat) produced in a resistance by a dc voltage is compared to that produced in the same resistance by an ac voltage of the same peak amplitude.*

A33.   *The effective value.*

A34.

$$I_{eff} = 0.707 \times I_{max}$$

A35.

$$E_{eff} = 0.707 \times E_{max}$$
$$= 0.707 \times E_{max}$$
$$= 0.707 \times 1{,}000 \text{ volts}$$
$$E_{eff} = 707 \text{ volts}$$

A36.

$$I_{max} = 1.414 \times I_{eff}$$

$$= 1.414 \times 4.25 \text{ amperes}$$

$$= 6 \text{ amperes.}$$

(Remember: Unless specified otherwise, the voltage or current value is always considered to be the effective value.)

A37. *When the two waves go through their maximum and minimum points at the same time and in the same direction.*

A38. *When the waves do not go through their maximum and minimum points at the same time, a PHASE DIFFERENCE exists, and the two waves are said to be out of phase. (Two waves are also considered to be out of phase if they differ in phase by 180° and their instantaneous voltages are always of opposite polarity, even though both waves go through their maximum and minimum points at the same time).*

A39. *They are in phase with each other.*

A40. *Locate the points on the time axis where the two waves cross traveling in the same direction. The number of degrees between these two points is the phase difference.*

A41.

$$I_{eff} = \frac{100}{45} = 2.22 \text{ ampers}$$

A42. *$I_{avg} = 0.636 \times I_{max} = 1.41$ amperes.*

A43. *43.3 ohms.*

CHAPTER 2

INDUCTANCE

LEARNING OBJECTIVES

Upon completion of this chapter you will be able to:

1. Write the basic unit of and the symbol for inductance.

2. State the type of moving field used to generate an emf in a conductor.

3. Define the term "inductance."

4. State the meanings of the terms "induced emf" and "counter emf."

5. State Lenz's law.

6. State the effect that inductance has on steady direct current, and direct current that is changing in magnitude.

7. List five factors that affect the inductance of a coil, and state how various physical changes in these factors affect inductance.

8. State the principles and sequences involved in the buildup and decay of current in an LR series circuit.

9. Write the formula for computing one time constant in an LR series circuit.

10. Solve L/R time constant problems.

11. State the three types of power loss in an inductor.

12. Define the term "mutual inductance."

13. State the meaning of the term "coupled circuits."

14. State the meaning of the term "coefficient of coupling."

15. Given the inductance values of and the coefficient of coupling between two series-connected inductors, solve for mutual inductance, M.

16. Write the formula for the "total inductance" of two inductors connected in series-opposing.

17. Given the inductance values of and the mutual inductance value between two coils connected in series-aiding, solve for their combined inductance, L_T.

INDUCTANCE

The study of inductance presents a very challenging but rewarding segment of electricity. It is challenging in the sense that, at first, it will seem that new concepts are being introduced. You will realize as this chapter progresses that these "new concepts" are merely extensions and enlargements of fundamental principles that you learned previously in the study of magnetism and electron physics. The study of inductance is rewarding in the sense that a thorough understanding of it will enable you to acquire a working knowledge of electrical circuits more rapidly.

CHARACTERISTICS OF INDUCTANCE

Inductance is the characteristic of an electrical circuit that opposes the starting, stopping, or a change in value of current. The above statement is of such importance to the study of inductance that it bears repeating. Inductance is the characteristic of an electrical conductor that OPPOSES CHANGE in CURRENT. The symbol for inductance is L and the basic unit of inductance is the HENRY (H). One henry is equal to the inductance required to induce one volt in an inductor by a change of current of one ampere per second.

You do not have to look far to find a physical analogy of inductance. Anyone who has ever had to push a heavy load (wheelbarrow, car, etc.) is aware that it takes more work to start the load moving than it does to keep it moving. Once the load is moving, it is easier to keep the load moving than to stop it again. This is because the load possesses the property of INERTIA. Inertia is the characteristic of mass which opposes a CHANGE in velocity. Inductance has the same effect on current in an electrical circuit as inertia has on the movement of a mechanical object. It requires more energy to start or stop current than it does to keep it flowing.

Q1. What is the basic unit of inductance and the abbreviation for this unit?

ELECTROMOTIVE FORCE (EMF)

You have learned that an electromotive force is developed whenever there is relative motion between a magnetic field and a conductor.

Electromotive force is a difference of potential or voltage which exists between two points in an electrical circuit. In generators and inductors the emf is developed by the action between the magnetic field and the electrons in a conductor. This is shown in figure 2-1.

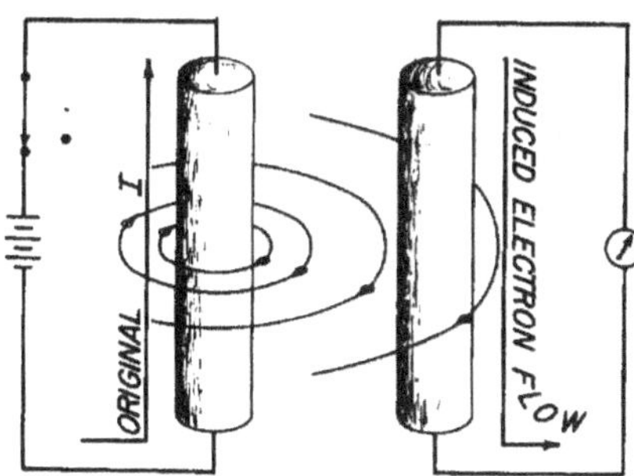

Figure 2-1.—Generation of an emf in an electrical conductor.

When a magnetic field moves through a stationary metallic conductor, electrons are dislodged from their orbits. The electrons move in a direction determined by the movement of the magnetic lines of flux. This is shown below:

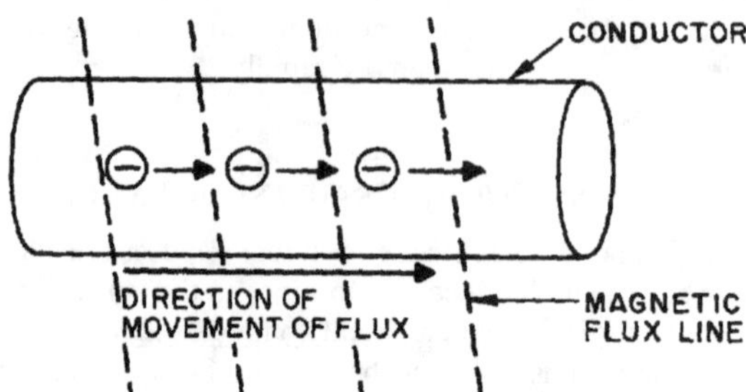

The electrons move from one area of the conductor into another area. The area that the electrons moved from has fewer negative charges (electrons) and becomes positively charged. The area the electrons move into becomes negatively charged. This is shown below:

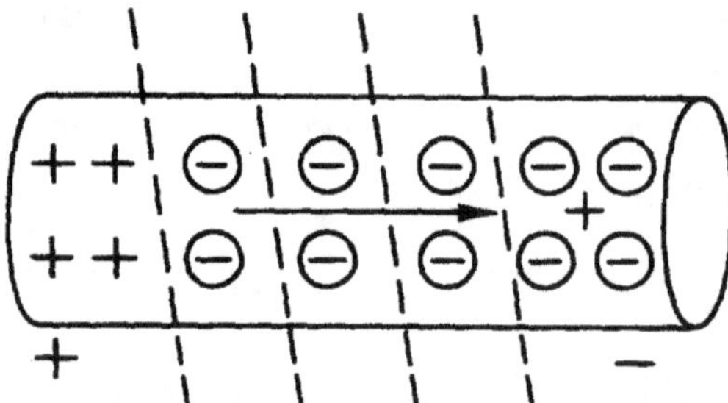

The difference between the charges in the conductor is equal to a difference of potential (or voltage). This voltage caused by the moving magnetic field is called electromotive force (emf).

In simple terms, the action of a moving magnetic field on a conductor can be compared to the action of a broom. Consider the moving magnetic field to be a moving broom. As the magnetic broom moves along (through) the conductor, it gathers up and pushes electrons before it, as shown below:

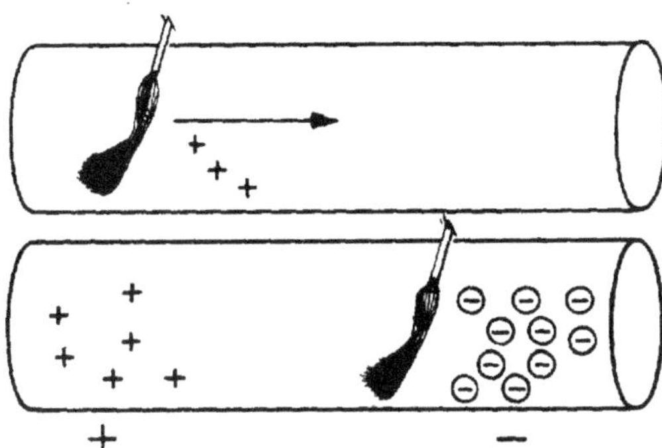

The area from which electrons are moved becomes positively charged, while the area into which electrons are moved becomes negatively charged. The potential difference between these two areas is the electromotive force or emf.

Q2. An emf is generated in a conductor when the conductor is cut by what type of field?

SELF-INDUCTANCE

Even a perfectly straight length of conductor has some inductance. As you know, current in a conductor produces a magnetic field surrounding the conductor. When the current changes, the magnetic field changes. This causes relative motion between the magnetic field and the conductor, and an electromotive force (emf) is induced in the conductor. This emf is called a SELF-INDUCED EMF because it is induced in the conductor carrying the current. The emf produced by this moving magnetic field is also referred to as COUNTER ELECTROMOTIVE FORCE (cemf). The polarity of the counter electromotive force is in the opposite direction to the applied voltage of the conductor. The overall effect will be to oppose a change in current magnitude. This effect is summarized by Lenz's law which states that: THE INDUCED EMF IN ANY CIRCUIT IS ALWAYS IN A DIRECTION TO OPPOSE THE EFFECT THAT PRODUCED IT.

If the shape of the conductor is changed to form a loop, then the electromagnetic field around each portion of the conductor cuts across some other portion of the same conductor. This is shown in its simplest form in figure 2-2. A length of conductor is looped so that two portions of the conductor lie next to each other. These portions are labeled conductor 1 and conductor 2. When the switch is closed, current (electron flow) in the conductor produces a magnetic field around ALL portions of the conductor. For simplicity, the magnetic field (expanding lines of flux) is shown in a single plane that is perpendicular to both conductors. Although the expanding field of flux originates at the same time in both conductors, it is considered as originating in conductor 1 and its effect on conductor 2 will be explained. With increasing current, the flux field expands outward from conductor 1, cutting across a portion of conductor 2. This results in an induced emf in conductor 2 as shown by the dashed arrow. Note that the induced emf is in the opposite direction to (in OPPOSITION to) the battery current and voltage, as stated in Lenz's law.

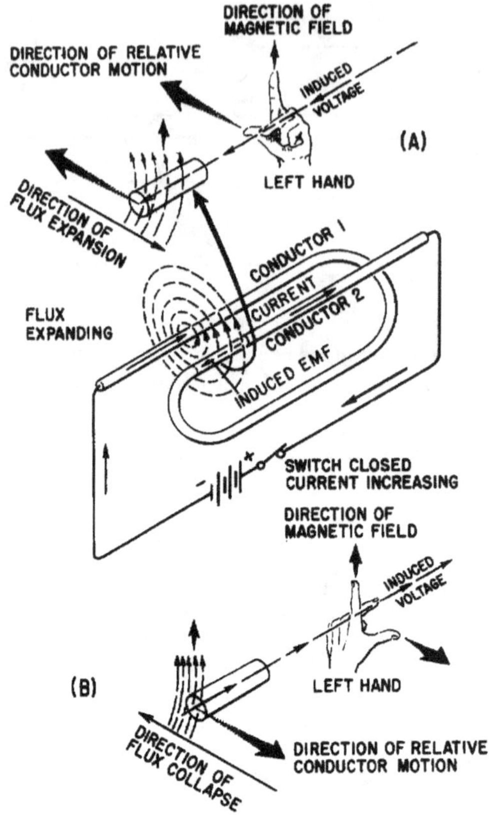

Figure 2-2.—Self-inductance.

The direction of this induced voltage may be determined by applying the LEFT-HAND RULE FOR GENERATORS. This rule is applied to a portion of conductor 2 that is "lifted" and enlarged for this purpose in figure 2-2(A). This rule states that if you point the thumb of your left hand in the direction of relative motion of the conductor and your index finger in the direction of the magnetic field, your middle finger, extended as shown, will now indicate the direction of the induced current which will generate the induced voltage (cemf) as shown.

In figure 2-2(B), the same section of conductor 2 is shown after the switch has been opened. The flux field is collapsing. Applying the left-hand rule in this case shows that the reversal of flux MOVEMENT has caused a reversal in the direction of the induced voltage. The induced voltage is now in the same direction as the battery voltage. The most important thing for you to note is that the self-induced voltage opposes BOTH changes in current. That is, when the switch is closed, this voltage delays the initial buildup of current by opposing the battery voltage. When the switch is opened, it keeps the current flowing in the same direction by aiding the battery voltage.

Thus, from the above explanation, you can see that when a current is building up it produces an expanding magnetic field. This field induces an emf in the direction opposite to the actual flow of current.

This induced emf opposes the growth of the current and the growth of the magnetic field. If the increasing current had not set up a magnetic field, there would have been no opposition to its growth. The whole reaction, or opposition, is caused by the creation or collapse of the magnetic field, the lines of which as they expand or contract cut across the conductor and develop the counter emf.

Since all circuits have conductors in them, you can assume that all circuits have inductance. However, inductance has its greatest effect only when there is a change in current. Inductance does NOT oppose current, only a CHANGE in current. Where current is constantly changing as in an ac circuit, inductance has more effect.

Q3. Define inductance.

Q4. What is meant by induced emf? By counter emf?

Q5. State Lenz's law.

Q6. What effect does inductance have (a) on steady direct current and (b) on direct current while it is changing in amplitude?

To increase the property of inductance, the conductor can be formed into a loop or coil. A coil is also called an inductor. Figure 2-3 shows a conductor formed into a coil. Current through one loop produces a magnetic field that encircles the loop in the direction as shown in figure 2-3(A). As current increases, the magnetic field expands and cuts all the loops as shown in figure 2-3(B). The current in each loop affects all other loops. The field cutting the other loop has the effect of increasing the opposition to a current change.

Figure 2-3.—Inductance.

Inductors are classified according to core type. The core is the center of the inductor just as the core of an apple is the center of an apple. The inductor is made by forming a coil of wire around a core. The core material is normally one of two basic types: soft-iron or air. An iron-core inductor and its schematic symbol (which is represented with lines across the top of it to indicate the presence of an iron core) are shown, in figure 2-4(A). The air-core inductor may be nothing more than a coil of wire, but it is usually a coil formed around a hollow form of some nonmagnetic material such as cardboard. This material serves no purpose other than to hold the shape of the coil. An air-core inductor and its schematic symbol are shown in figure 2-4(B).

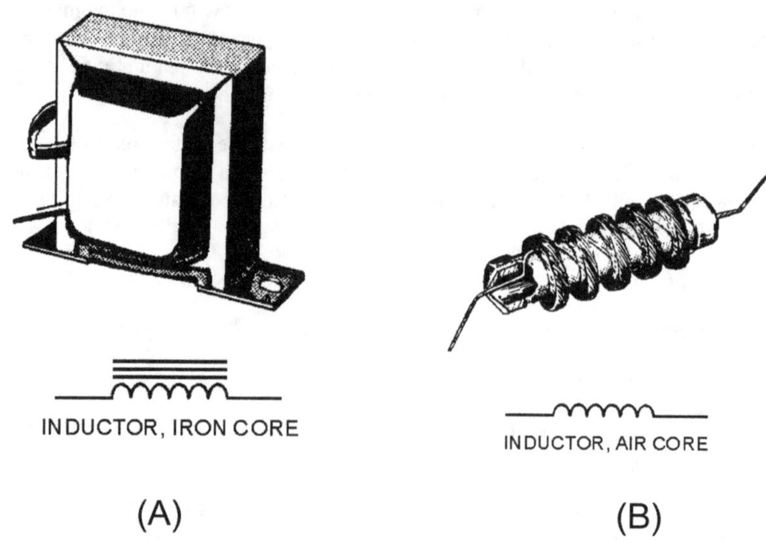

Figure 2-4.—Inductor types and schematic symbols.

Factors Affecting Coil Inductance

There are several physical factors which affect the inductance of a coil. They include the number of turns in the coil, the diameter of the coil, the coil length, the type of material used in the core, and the number of layers of winding in the coils.

Inductance depends entirely upon the physical construction of the circuit, and can only be measured with special laboratory instruments. Of the factors mentioned, consider first how the number of turns affects the inductance of a coil. Figure 2-5 shows two coils. Coil (A) has two turns and coil (B) has four turns. In coil (A), the flux field set up by one loop cuts one other loop. In coil (B), the flux field set up by one loop cuts three other loops. Doubling the number of turns in the coil will produce a field twice as strong, if the same current is used. A field twice as strong, cutting twice the number of turns, will induce four times the voltage. Therefore, it can be said that the <u>inductance varies as the square of the number of turns</u>.

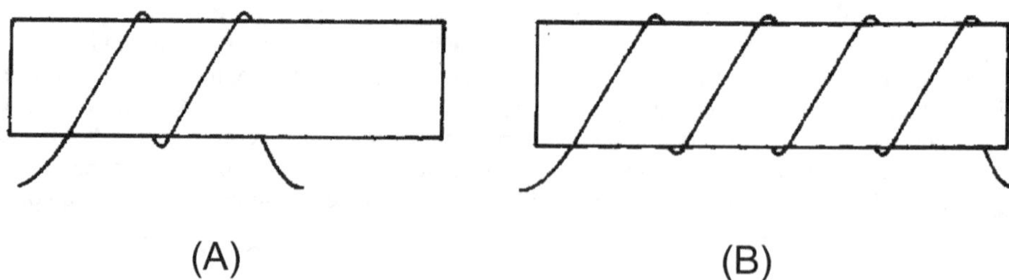

Figure 2-5.—Inductance factor (turns).

The second factor is the coil diameter. In figure 2-6 you can see that the coil in view B has twice the diameter of coil view A. Physically, it requires more wire to construct a coil of large diameter than one of small diameter with an equal number of turns. Therefore, more lines of force exist to induce a counter emf in the coil with the larger diameter. Actually, the inductance of a coil <u>increases directly as the cross-sectional area of the core increases</u>. Recall the formula for the area of a circle: $A = \pi r^2$. Doubling the radius of a coil increases the inductance by a factor of four.

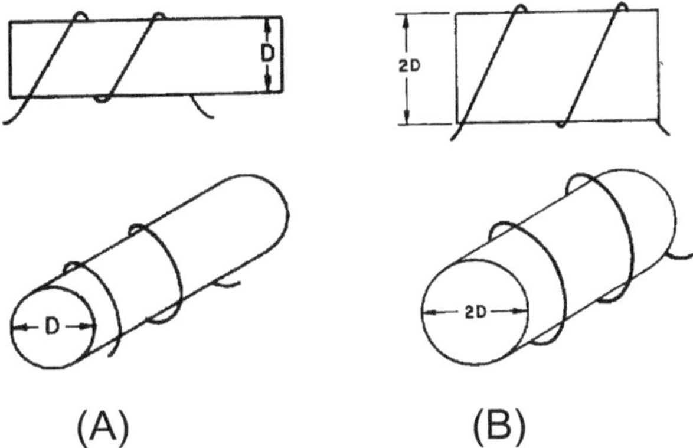

(A)   (B)

Figure 2-6.—Inductance factor (diameter).

The third factor that affects the inductance of a coil is the length of the coil. Figure 2-7 shows two examples of coil spacings. Coil (A) has three turns, rather widely spaced, making a relatively long coil. A coil of this type has few flux linkages, due to the greater distance between each turn. Therefore, coil (A) has a relatively low inductance. Coil (B) has closely spaced turns, making a relatively short coil. This close spacing increases the flux linkage, increasing the inductance of the coil. <u>Doubling the length of a coil while keeping the same number of turns halves the value of inductance.</u>

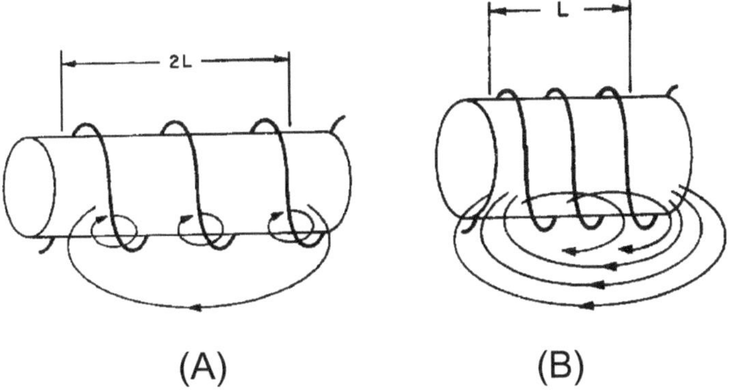

(A)   (B)

Figure 2-7.—Inductance factor (coil length). CLOSELY WOUND

The fourth physical factor is the type of core material used with the coil. Figure 2-8 shows two coils: Coil (A) with an air core, and coil (B) with a soft-iron core. The magnetic core of coil (B) is a better path for magnetic lines of force than is the nonmagnetic core of coil (A). The soft-iron magnetic core's high permeability has less reluctance to the magnetic flux, resulting in more magnetic lines of force. This increase in the magnetic lines of force increases the number of lines of force cutting each loop of the coil, thus increasing the inductance of the coil. It should now be apparent that the <u>inductance of a coil increases directly as the permeability of the core material increases</u>.

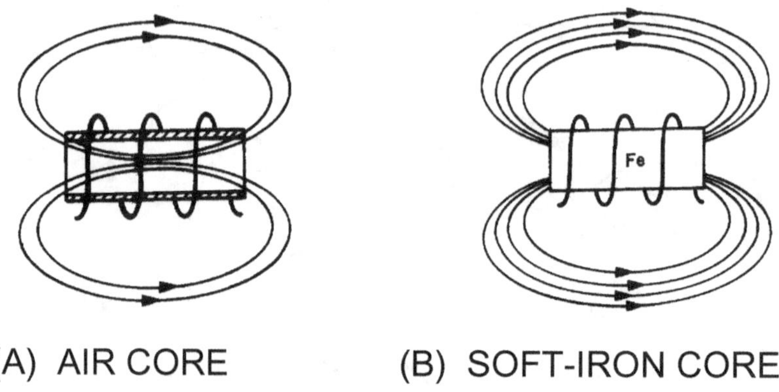

Figure 2-8.—Inductance factor (core material).

Another way of increasing the inductance is to wind the coil in layers. Figure 2-9 shows three cores with different amounts of layering. The coil in figure 2-9(A) is a poor inductor compared to the others in the figure because its turns are widely spaced and there is no layering. The flux movement, indicated by the dashed arrows, does not link effectively because there is only one layer of turns. A more inductive coil is shown in figure 2-9(B). The turns are closely spaced and the wire has been wound in two layers. The two layers link each other with a greater number of flux loops during all flux movements. Note that nearly all the turns, such as X, are next to four other turns (shaded). This causes the flux linkage to be increased.

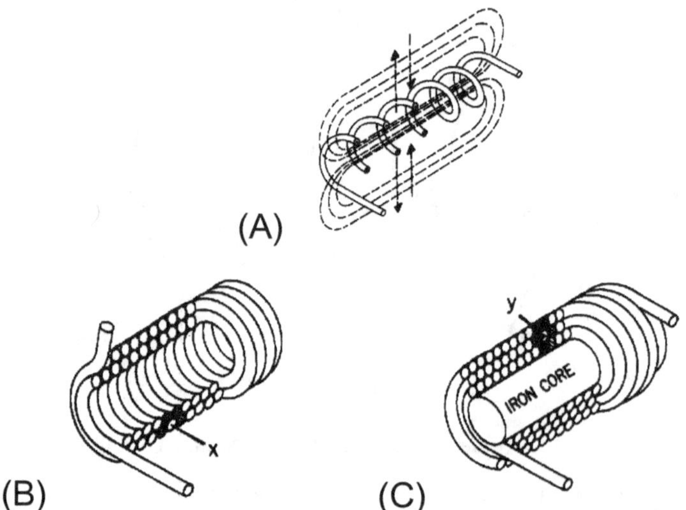

Figure 2-9.—Coils of various inductances.

A coil can be made still more inductive by winding it in three layers, as shown in figure 2-9(C). The increased number of layers (cross-sectional area) improves flux linkage even more. Note that some turns, such as Y, lie directly next to six other turns (shaded). In actual practice, layering can continue on through many more layers. The important fact to remember, however, is that <u>the inductance of the coil increases with each layer added</u>.

As you have seen, several factors can affect the inductance of a coil, and all of these factors are variable. Many differently constructed coils can have the same inductance. The important information to remember, however, is that <u>inductance is dependent upon the degree of linkage between the wire conductor(s) and the electromagnetic field</u>. In a straight length of conductor, there is very little flux linkage between one part of the conductor and another. Therefore, its inductance is extremely small. It was shown that conductors become much more inductive when they are wound into coils. This is true because there is maximum flux linkage between the conductor turns, which lie side by side in the coil.

Q7.

    a. List five factors that affect the inductance of a coil.

    b. Bending a straight piece of wire into a loop or coil has what effect on the inductance of the wire?

    c. Doubling the number of turns in a coil has what effect on the inductance of the coil?

    d. Decreasing the diameter of a coil has what effect on the inductance of the coil?

    e. Inserting a soft-iron core into a coil has what effect on the inductance of the coil?

    f. Increasing the number of layers of windings in a coil has what effect on the inductance of the coil?

UNIT OF INDUCTANCE

As stated before, the basic unit of inductance (L) is the HENRY (H), named after Joseph Henry, the co-discoverer with Faraday of the principle of electromagnetic induction. <u>An inductor has an inductance of 1 henry if an emf of 1 volt is induced in the inductor when the current through the inductor is changing at the rate of 1 ampere per second</u>. The relationship between the induced voltage, the inductance, and the rate of change of current with respect to time is stated mathematically as:

$$E_{ind} = L\frac{\Delta I}{\Delta t}$$

where E_{ind} is the induced emf in volts; L is the inductance in henrys; and ΔI is the change in current in amperes occurring in Δt seconds. The symbol Δ (Greek letter delta), means "a change in ....". The henry is a large unit of inductance and is used with relatively large inductors. With small inductors, the millihenry is used. (A millihenry is equal to 1×10^{-3} henry, and one henry is equal to 1,000 millihenrys.) For still smaller inductors the unit of inductance is the microhenry (μH). (A μH = 1×10^{-6}H, and one henry is equal to 1,000,000 microhenrys.)

GROWTH AND DECAY OF CURRENT IN AN LR SERIES CIRCUIT

When a battery is connected across a "pure" inductance, the current builds up to its final value at a rate determined by the battery voltage and the internal resistance of the battery. The current buildup is gradual because of the counter emf generated by the self-inductance of the coil. When the current starts to flow, the magnetic lines of force move outward from the coil. These lines cut the turns of wire on the inductor and build up a counter emf that opposes the emf of the battery. This opposition causes a delay in the time it takes the current to build up to a steady value. When the battery is disconnected, the lines of force collapse. Again these lines cut the turns of the inductor and build up an emf that tends to prolong the flow of current.

A voltage divider containing resistance and inductance may be connected in a circuit by means of a special switch, as shown in figure 2-10(A). Such a series arrangement is called an LR series circuit.

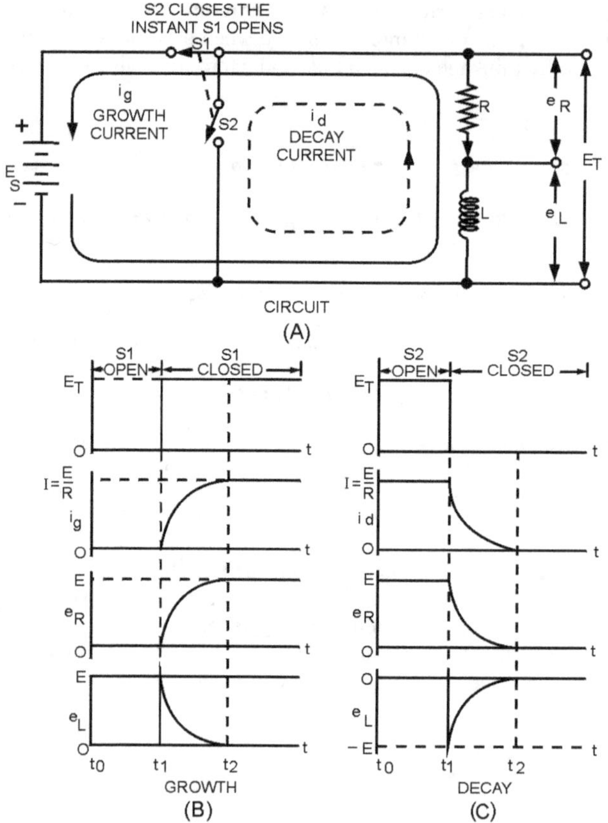

Figure 2-10.—Growth and decay of current in an LR series circuit.

When switch S_1 is closed (as shown), a voltage E_S appears across the voltage divider. At this instant the current will attempt to increase to its maximum value. However, this instantaneous current change causes coil L to produce a back EMF, which is opposite in polarity and almost equal to the EMF of the source. This back EMF opposes the rapid current change. Figure 2-10(B) shows that at the instant switch S_1 is closed, there is no measurable growth current (i_g), a minimum voltage drop is across resistor R, and maximum voltage exists across inductor L.

As current starts to flow, a voltage (e_R) appears across R, and the voltage across the inductor is reduced by the same amount. The fact that the voltage across the inductor (L) is reduced means that the growth current (i_g) is increased and consequently e_R is increased. Figure 2-10(B) shows that the voltage across the inductor (e_L) finally becomes zero when the growth current (i_g) stops increasing, while the voltage across the resistor (e_R) builds up to a value equal to the source voltage (E_S).

Electrical inductance is like mechanical inertia, and the growth of current in an inductive circuit can be likened to the acceleration of a boat on the surface of the water. The boat does not move at the instant a constant force is applied to it. At this instant all the applied force is used to overcome the inertia of the boat. Once the inertia is overcome the boat will start to move. After a while, the speed of the boat reaches its maximum value and the applied force is used up in overcoming the friction of the water against the hull.

When the battery switch (S_1) in the LR circuit of figure 2-10(A) is closed, the rate of the current increase is maximum in the inductive circuit. At this instant all the battery voltage is used in overcoming the emf of self-induction which is a maximum because the rate of change of current is maximum. Thus the battery voltage is equal to the drop across the inductor and the voltage across the resistor is zero. As time goes on more of the battery voltage appears across the resistor and less across the inductor. The rate of change of current is less and the induced emf is less. As the steady-state condition of the current is approached, the drop across the inductor approaches zero and all of the battery voltage is "dropped" across the resistance of the circuit.

Thus the voltages across the inductor and the resistor change in magnitude during the period of growth of current the same way the force applied to the boat divides itself between the effects of inertia and friction. In both examples, the force is developed first across the inertia/inductive effect and finally across the friction/resistive effect.

Figure 2-10(C) shows that when switch S_2 is closed (source voltage E_S removed from the circuit), the flux that has been established around the inductor (L) collapses through the windings. This induces a voltage e_L in the inductor that has a polarity opposite to E_S and is essentially equal to E_S in magnitude. The induced voltage causes decay current (i_d) to flow in resistor R in the same direction in which current was flowing originally (when S_1 was closed). A voltage (e_R) that is initially equal to source voltage (E_S) is developed across R. The voltage across the resistor (e_R) rapidly falls to zero as the voltage across the inductor (e_L) falls to zero due to the collapsing flux.

Just as the example of the boat was used to explain the growth of current in a circuit, it can also be used to explain the decay of current in a circuit. When the force applied to the boat is removed, the boat still continues to move through the water for a while, eventually coming to a stop. This is because energy was being stored in the inertia of the moving boat. After a period of time the friction of the water overcomes the inertia of the boat, and the boat stops moving. Just as inertia of the boat stored energy, the magnetic field of an inductor stores energy. Because of this, even when the power source is removed, the stored energy of the magnetic field of the inductor tends to keep current flowing in the circuit until the magnetic field collapse.

Q8.

    a.   When voltage is first applied to a series LR circuit, how much opposition does the inductance have to the flow of current compared to that of the circuit resistance?

    b.   In a series circuit containing a resistor (R_1) and an inductor (L_1), what voltage exists across R_1 when the counter emf is at its maximum value?

    c.   What happens to the voltage across the resistance in an LR circuit during current buildup in the circuit, and during current decay in the circuit?

L/R Time Constant

The L/R TIME CONSTANT is a valuable tool for use in determining the time required for current in an inductor to reach a specific value. As shown in figure 2-11, one L/R time constant is the time required for the current in an inductor to increase to 63 percent (actually 63.2 percent) of the maximum current. Each time constant is equal to the time required for the current to increase by 63.2 percent of the difference in value between the current flowing in the inductor and the maximum current. Maximum current flows in the inductor after five L/R time constants are completed. The following example should clear up any confusion about time constants. Assume that maximum current in an LR circuit is 10 amperes. As you know, when the circuit is energized, it takes time for the current to go from zero to 10 amperes. When the first time constant is completed, the current in the circuit is equal to 63.2% of 10 amperes. Thus the amplitude of current at the end of 1 time constant is 6.32 amperes.

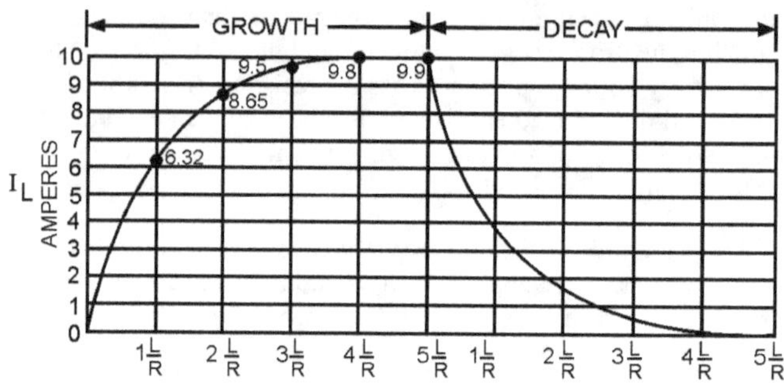

Figure 2-11.—L/R time constant.

During the second time constant, current again increases by 63.2% (.632) of the difference in value between the current flowing in the inductor and the maximum current. This difference is 10 amperes minus 6.32 amperes and equals 3.68 amperes; 63.2% of 3.68 amperes is 2.32 amperes. This increase in current during the second time constant is added to that of the first time constant. Thus, upon completion of the second time constant, the amount of current in the LR circuit is 6.32 amperes + 2.32 amperes = 8.64 amperes.

During the third constant, current again increases:

$$10 \text{ amperes} - 8.64 \text{ amperes} = 1.36 \text{ amperes}$$
$$1.36 \text{ amperes} \times .632 = 0.860 \text{ ampere}$$
$$8.64 \text{ amperes} + 0.860 \text{ ampere} = 9.50 \text{ amperes}$$

During the fourth time constant, current again increases:

$$10 \text{ amperes} - 9.50 \text{ amperes} = 0.5 \text{ ampere}$$
$$0.5 \text{ ampere} \times .632 = 0.316 \text{ ampere}$$
$$9.50 \text{ amperes} + 0.316 \text{ ampere} = 9.82 \text{ amperes}$$

During the fifth time constant, current increases as before:

$$10 \text{ amperes} - 9.82 \text{ amperes} = 0.18 \text{ ampere}$$

$$0.18 \text{ ampere} \times .632 = 0.114 \text{ ampere}$$

$$9.82 \text{ amperes} + .114 \text{ ampere} = 9.93 \text{ amperes}$$

Thus, the current at the end of the fifth time constant is almost equal to 10.0 amperes, the maximum current. For all practical purposes the slight difference in value can be ignored.

When an LR circuit is deenergized, the circuit current decreases (decays) to zero in five time constants at the same rate that it previously increased. If the growth and decay of current in an LR circuit are plotted on a graph, the curve appears as shown in figure 2-11. Notice that current increases and decays at the same rate in five time constants.

The value of the time constant in seconds is equal to the inductance in henrys divided by the circuit resistance in ohms.

The formula used to calculate one L/R time constant is:

$$\text{Time Constant (TC) in seconds} = \frac{L \text{ (in henrys)}}{R \text{ (in ohms)}}$$

Q9.   What is the formula for one L/R time constant?

Q10.

    a.   The maximum current applied to an inductor is 1.8 amperes. How much current flowed in the inductor 3 time constants after the circuit was first energized?

    b.   What is the minimum number of time constants required for the current in an LR circuit to increase to its maximum value?

    c.   A circuit containing only an inductor and a resistor has a maximum of 12 amperes of applied current flowing in it. After 5 L/R time constants the circuit is opened. How many time constants is required for the current to decay to 1.625 amperes?

POWER LOSS IN AN INDUCTOR

Since an inductor (coil) consists of a number of turns of wire, and since all wire has some resistance, every inductor has a certain amount of resistance. Normally this resistance is small. It is usually neglected in solving various types of ac circuit problems because the reactance of the inductor (the opposition to alternating current, which will be discussed later) is so much greater than the resistance that the resistance has a negligible effect on the current.

However, since some inductors are designed to carry relatively large amounts of current, considerable power can be dissipated in the inductor even though the amount of resistance in the inductor is small. This power is wasted power and is called COPPER LOSS. The copper loss of an inductor can be calculated by multiplying the square of the current in the inductor by the resistance of the winding (I^2R).

In addition to copper loss, an iron-core coil (inductor) has two iron losses. These are called HYSTERESIS LOSS and EDDY-CURRENT LOSS. Hysteresis loss is due to power that is consumed in reversing the magnetic field of the inductor core each time the direction of current in the inductor changes.

Eddy-current loss is due to heating of the core by circulating currents that are induced in the iron core by the magnetic field around the turns of the coil. These currents are called eddy currents and circulate within the iron core only.

All these losses dissipate power in the form of heat. Since this power cannot be returned to the electrical circuit, it is lost power.

Q11. State three types of power loss in an inductor.

MUTUAL INDUCTANCE

Whenever two coils are located so that the flux from one coil links with the turns of the other coil, a change of flux in one coil causes an emf to be induced in the other coil. This allows the energy from one coil to be transferred or coupled to the other coil. The two coils are said to be coupled or linked by the property of MUTUAL INDUCTANCE (M). The amount of mutual inductance depends on the relative positions of the two coils. This is shown in figure 2-12. If the coils are separated a considerable distance, the amount of flux common to both coils is small and the mutual inductance is low. Conversely, if the coils are close together so that nearly all the flux of one coil links the turns of the other, the mutual inductance is high. The mutual inductance can be increased greatly by mounting the coils on a common iron core.

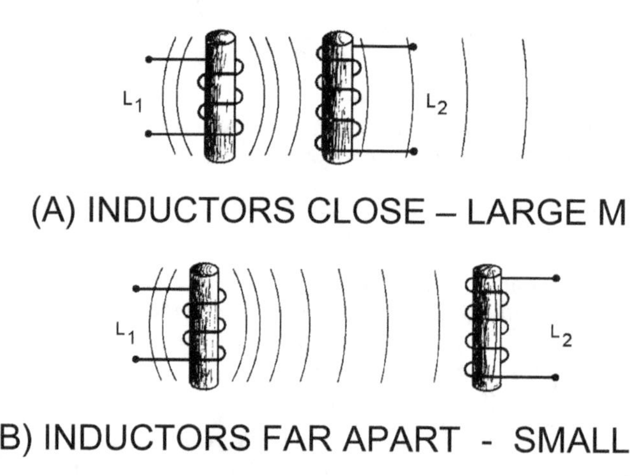

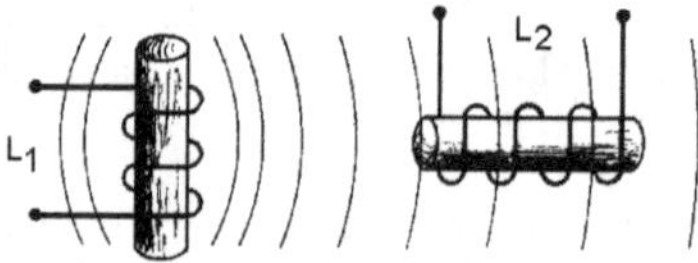

Figure 2-12.—The effect of position of coils on mutual inductance (M).

Two coils are placed close together as shown in figure 2-13. Coil 1 is connected to a battery through switch S, and coil 2 is connected to an ammeter (A). When switch S is closed as in figure 2-13(A), the current that flows in coil 1 sets up a magnetic field that links with coil 2, causing an induced voltage in coil 2 and a momentary deflection of the ammeter. When the current in coil 1 reaches a steady value, the ammeter returns to zero. If switch S is now opened as in figure 2-13(B), the ammeter (A) deflects momentarily in the opposite direction, indicating a momentary flow of current in the opposite direction in coil 2. This current in coil 2 is produced by the collapsing magnetic field of coil 1.

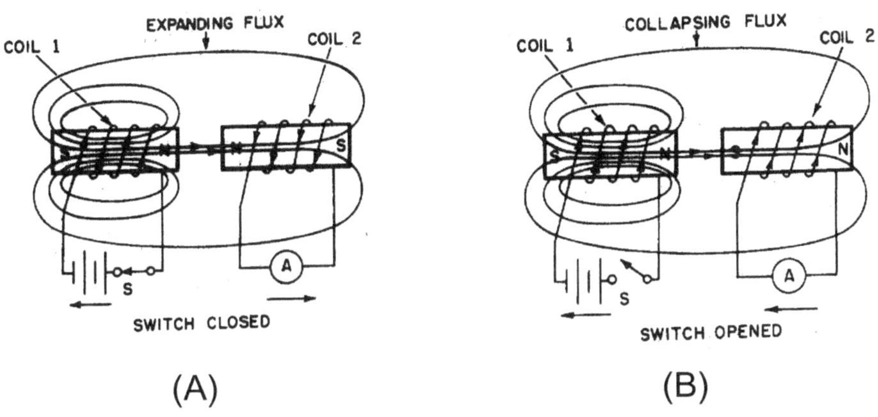

Figure 2-13.—Mutual inductance.

Q12.   Define mutual inductance.

FACTORS AFFECTING MUTUAL INDUCTANCE

The mutual inductance of two adjacent coils is dependent upon the physical dimensions of the two coils, the number of turns in each coil, the distance between the two coils, the relative positions of the axes of the two coils, and the permeability of the cores.

The COEFFICIENT OF COUPLING between two coils is equal to the ratio of the flux cutting one coil to the flux originated in the other coil. If the two coils are so positioned with respect to each other so that all of the flux of one coil cuts all of the turns of the other, the coils are said to have a unity coefficient of coupling. It is never exactly equal to unity (1), but it approaches this value in certain types of coupling devices. If all of the flux produced by one coil cuts only half the turns of the other coil, the coefficient of coupling is 0.5. The coefficient of coupling is designated by the letter K.

The mutual inductance between two coils, L_1 and L_2, is expressed in terms of the inductance of each coil and the coefficient of coupling K. As a formula:

$$M = K\sqrt{L_1 L_2}$$

$$
\begin{aligned}
\text{where:} \quad M &= \text{Mutual inductance in henrys} \\
K &= \text{Coefficient of coupling} \\
L_1, L_2 &= \text{Inductance of coil in henrys}
\end{aligned}
$$

Example problem:

One 10-H coil and one 20-H coil are connected in series and are physically close enough to each other so that their coefficient of coupling is 0.5. What is the mutual inductance between the coils?

$$
\begin{aligned}
\text{Use the formula:} \quad M &= K\sqrt{L_1 L_2} \\
M &= 0.5\sqrt{(10H)(20H)} \\
M &= 0.5\sqrt{200H} \\
M &= 0.5 \times 14.14H \\
M &= 7.07H
\end{aligned}
$$

Q13. When are two circuits said to be coupled?

Q14. What is meant by the coefficient of coupling?

SERIES INDUCTORS WITHOUT MAGNETIC COUPLING

When inductors are well shielded or are located far enough apart from one another, the effect of mutual inductance is negligible. If there is no mutual inductance (magnetic coupling) and the inductors are connected in series, the total inductance is equal to the sum of the individual inductances. As a formula:

$$L_T = L_1 + L_2 + L_3 + .. L_n$$

where L_T is the total inductance; L_1, L_2, L_3 are the inductances of L_1, L_2, L_3; and L_n means that any number (n) of inductors may be used. The inductances of inductors in series are added together like the resistances of resistors in series.

SERIES INDUCTORS WITH MAGNETIC COUPLING

When two inductors in series are so arranged that the field of one links the other, the combined inductance is determined as follows:

$$L_T = L_1 + L_2 \pm 2M$$

$$
\begin{aligned}
\text{where:} \quad L_T &= \text{The total inductance} \\
L_1, L_2 &= \text{The inductances of } L_1, L_2 \\
M &= \text{The mutual inductance between} \\
&\quad \text{the two inductors}
\end{aligned}
$$

The plus sign is used with M when the magnetic fields of the two inductors are aiding each other, as shown in figure 2-14. The minus sign is used with M when the magnetic field of the two inductors oppose each other, as shown in figure 2-15. The factor 2M accounts for the influence of L_1 on L_2 and L_2 on L_1.

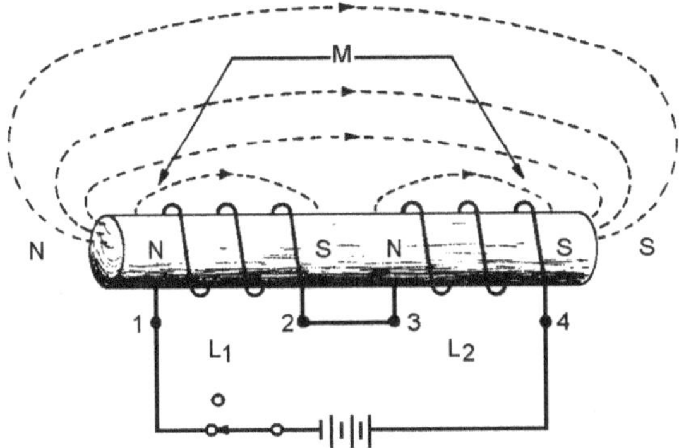

Figure 2-14.—Series inductors with aiding fields.

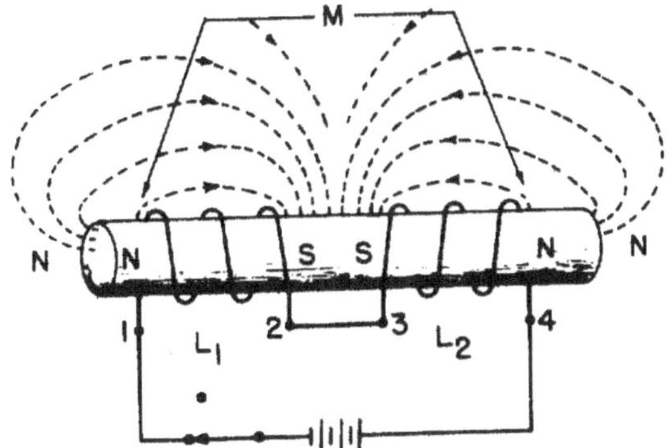

Figure 2-15.—Series inductors with opposing fields.

Example problem:

A 10-H coil is connected in series with a 5-H coil so the fields aid each other. Their mutual inductance is 7 H. What is the combined inductance of the coils?

$$\text{Use the formula:} \quad L_T = L_1 + L_2 + 2M$$
$$L_T = 10\,H + 5\,H + 2(7\,H)$$
$$L_T = 29\,H$$

Q15. *Two series-connected 7-H inductors are adjacent to each other; their coefficient of coupling is 0.64. What is the value of M?*

Q16. *A circuit contains two series inductors aligned in such a way that their magnetic fields oppose each other. What formula should you use to compute total inductance in this circuit?*

Q17. *The magnetic fields of two coils are aiding each other. The inductance of the coils are 3 H and 5 H, respectively. The coils' mutual inductance is 5 H. What is their combined inductance?*

PARALLEL INDUCTORS WITHOUT COUPLING

The total inductance (L_T) of inductors in parallel is calculated in the same manner that the total resistance of resistors in parallel is calculated, provided the coefficient of coupling between the coils is zero. Expressed mathematically:

$$\frac{1}{L_T} = \frac{1}{L_1} + \frac{1}{L_2} + \frac{1}{L_3} \dots + \frac{1}{L_N}$$

SUMMARY

The important points of this chapter are summarized below. Study this information before continuing, as this information will lay the foundation for later chapters.

INDUCTANCE—The characteristic of an electrical circuit that opposes a change in current. The reaction (opposition) is caused by the creation or destruction of a magnetic field. When current starts to flow, magnetic lines of force are created. These lines of force cut the conductor inducing a counter emf in a <u>direction that opposes current</u>.

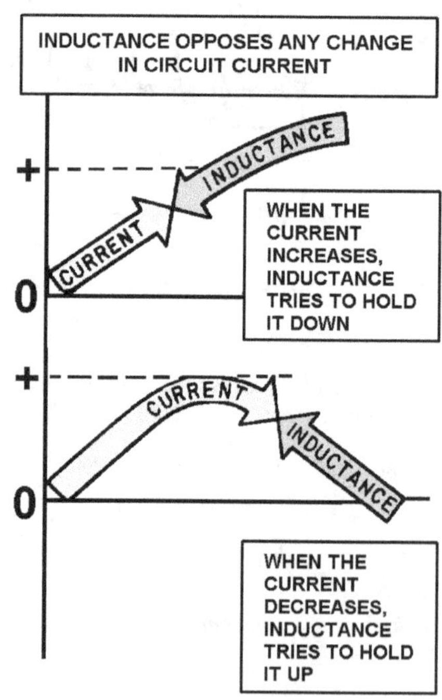

SELF-INDUCTANCE—The process by which a circuit induces an emf into itself by its own moving magnetic field. All electrical circuits possess self-inductance. This opposition (inductance), however, only takes place when there is a change in current. Inductance does NOT oppose current, only a CHANGE in current. The property of inductance can be increased by forming the conductor into a loop. In a loop, the magnetic lines of force affect more of the conductor at one time. This increases the self-induced emf.

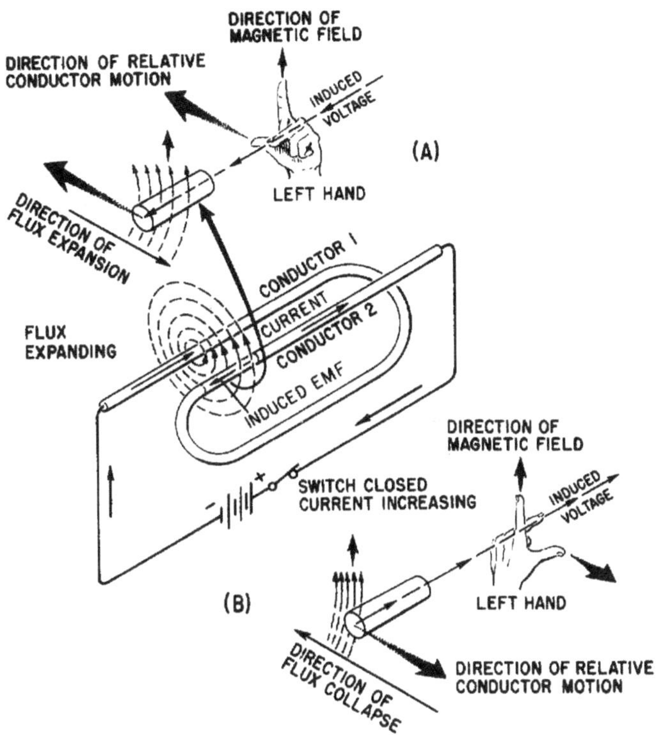

INDUCTANCE OF A COIL—The property of inductance can be further increased if the conductor is formed into a coil. Because a coil contains more loops, more of the conductor can be affected by the magnetic field. Inductors (coils) are classified according to core type. The core material is normally either air or soft iron.

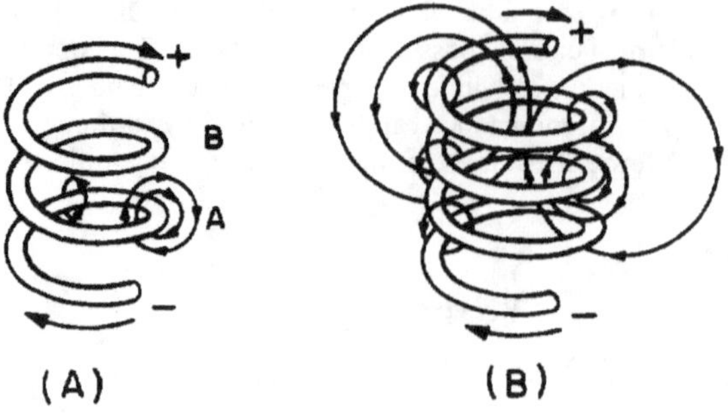

<u>FACTORS AFFECTING COIL INDUCTANCE</u>—The inductance of a coil is entirely dependent upon its physical construction. Some of the factors affecting the inductance are:

- The number of turns in the coil. Increasing the number of turns will increase the inductance.

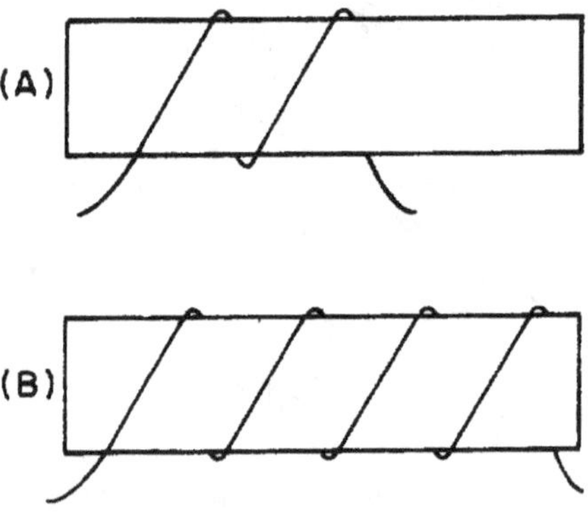

- The coil diameter. The inductance increases directly as the cross-sectional area of the coil increases.

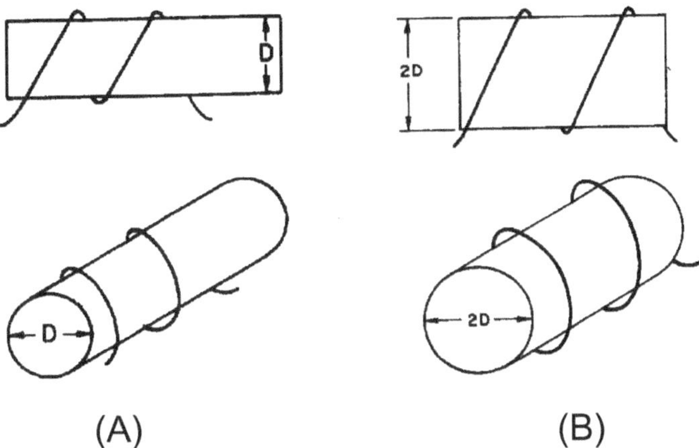

(A)   (B)

- The length of the coil. When the length of the coil is increased while keeping the number of turns the same, the turn-spacing is increased. This decreases the inductance of the coil.

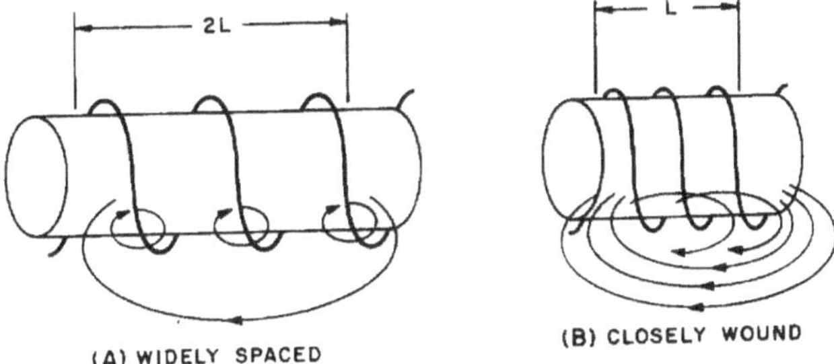

- The type of core material. Increasing the permeability of the core results in increasing the inductance of the coil.

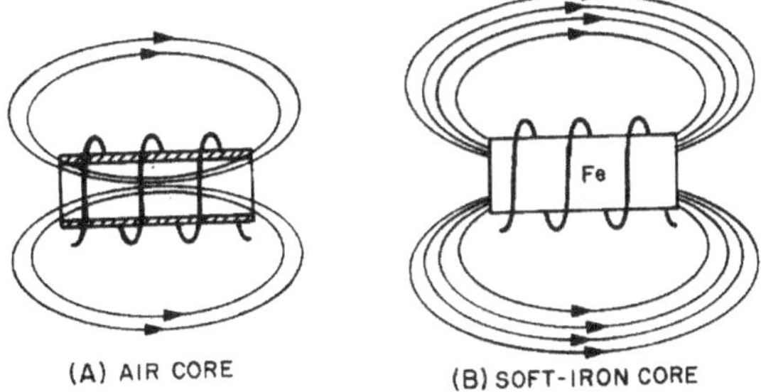

- Winding the coil in layers. The more layers used to form a coil, the greater effect the magnetic field has on the conductor. By layering a coil, you can increase the inductance.

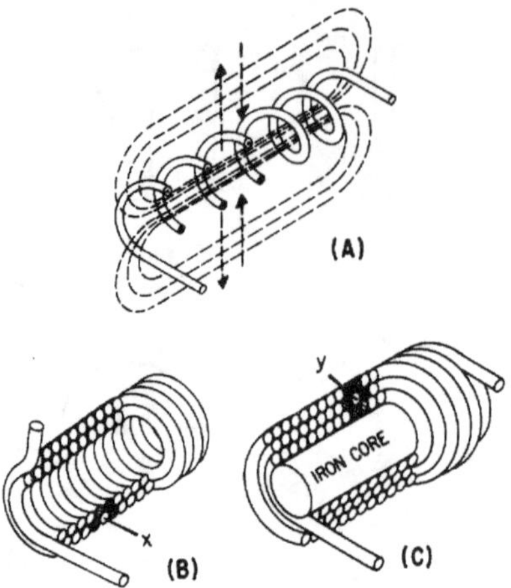

UNIT OF INDUCTANCE—Inductance (L) is measured in henrys (H). An inductor has an inductance of one henry (H) if an emf of one volt is induced in the inductor when the current through the inductor is changing at the rate of 1 ampere per second. Common units of inductance are henry (H), millihenry (mH) and the microhenry (μH).

GROWTH AND DECAY OF CURRENT IN AN LR CIRCUIT—The required for the current in an inductor to increase to 63.2 percent of the maximum current or to decrease to 36.8 percent of the maximum current is known as the time constant. The letter symbol for an LR time constant is L/R. As a formula:

$$\frac{L}{R}.$$

As a formula:

$$t \text{ (in seconds)} = \frac{L \text{ (in henrys)}}{R \text{ (in ohms)}}$$

or

$$t \text{ (in microseconds)} = \frac{L \text{ (in microhenrys)}}{R \text{ (in ohms)}}$$

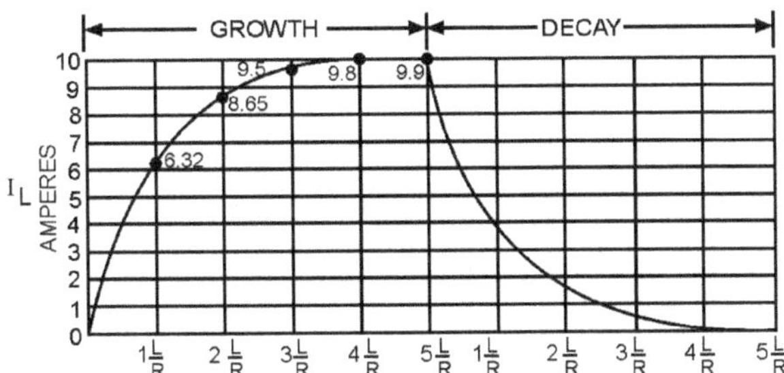

The time constant of an LR circuit may also be defined as the time required for the current in the inductor to grow or decay to its final value if it continued to grow or decay at its initial rate. For all practical purposes, the current in the inductor reaches a maximum value in 5 "Time Constants" and decreases to zero in 5 "Time Constants".

POWER LOSSES IN AN INDUCTOR—Since an inductor (coil) contains a number turns of wire, and all wire has some resistance, the inductor has a certain amount of resistance. This resistance is usually very small and has a negligible effect on current. However, there are power losses in an inductor. The main power losses in an inductor are copper loss, hysteresis loss, and eddy-current loss. Copper loss can be calculated by multiplying the square of the current by the resistance of the wire in the coil (I_2R). Hysteresis loss is due to power that is consumed in reversing the magnetic field of the core each time the current direction changes. Eddy-current loss is due to core heating caused by circulating currents induced in an iron core by the magnetic field of the coil.

MUTUAL INDUCTANCE—When two coils are located so that the flux from one coil cuts the turns of the other coil, the coils have mutual inductance. The amount of mutual inductance depends upon several factors: the relative position of the axes of the two coils; the permeability of the cores; the physical dimensions of the two coils; the number of turns in each coil, and the distance between the coils. The coefficient of coupling K specifies the amount of coupling between the coils. If all of the flux from one coil cuts all of the turns of the other coil, the coefficient of coupling K is 1 or unity. If none of the flux from one coil cuts the turns of the other coil, the coefficient of coupling is zero. The mutual inductance between two coils (L_1 and L_2) may be expressed mathematically as:

$$M = K\sqrt{L_1L_2}$$

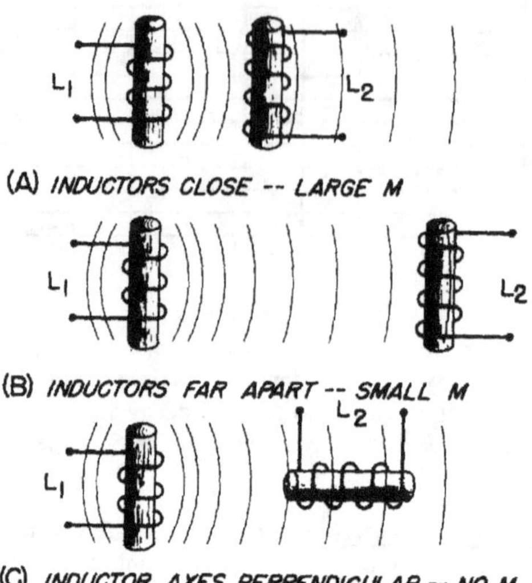

COMPUTING THE INDUCTANCE OF A CIRCUIT—When the total inductance of a circuit is computed, the individual inductive values are treated the same as resistance values. The inductances of inductors in series are added like the resistances of resistors in series. That is,

$$L_T = L_1 + L_2 + L_3 .... + L_n$$

The inductances of inductors in parallel are combined mathematically like the resistances of resistors in parallel. That is,

$$L_T = \cfrac{1}{\dfrac{1}{L_1} + \dfrac{1}{L_2} + \dfrac{1}{L_3} ... \dfrac{1}{L_n}}$$

Both of the above formulas are accurate, providing there is no mutual inductance between the inductors.

ANSWERS TO QUESTIONS Q1. THROUGH Q17.

A1.   *The henry, H.*

A2.   *Magnetic field.*

A3.   *Inductance is the property of a coil (or circuit) which opposes any CHANGE in current.*

A4.   *Induced emf is the emf which appears across a conductor when there is relative motion between the conductor and a magnetic field; counter emf is the emf induced in a conductor that opposes the applied voltage.*

A5.   *The induced emf in any circuit is in a direction to oppose the effect that produced it.*

A6.

    a.   *No effect.*

    b.   *Inductance opposes any change in the amplitude of current.*

A7.

    a.

        1.   *The numbers of turns in a coil.*

        2.   *The type of material used in the core.*

        3.   *The diameter of the coil.*

        4.   *The coil length.*

        5.   *The number of layers of windings in the coil.*

    b.   *Increases inductance.*

    c.   *Increases inductance.*

    d.   *Decreases inductance.*

    e.   *Increases inductance.*

    f.   *Increases inductance.*

A8.

    a.   *Inductance causes a very large opposition to the flow of current when voltage is first applied to an LR circuit; resistance causes comparatively little opposition to current at that time.*

    b.   *Zero.*

    c.   *During current buildup, the voltage across the resistor gradually increases to the same voltage as the source voltage; and during current decay the voltage across the resistor gradually drops to zero*

A9.

$$t = \frac{L}{R}$$

A10.

    a.   1.71 amperes.

    b.   5 time constants.

    c.   2 time constants.

A11.   Copper loss; hysteresis loss; eddy-current loss.

A12.   Mutual inductance is the property existing between two coils so positioned that flux from one coil cuts the windings of the other coil.

A13.   When they are arranged so that energy from one circuit is transferred to the other circuit.

A14.   The ratio of the fines of force produced by one coil to the lines of force that link another coil. It is never greater than one.

A15.

$$4.48\ \text{H} \left(\text{because } M = K \sqrt{7H \times 7H} = 0.64 \times 7H = 4.48H \right)$$

A16.

$$L_T = L_1 + L_2 - 2M$$

A17.

$$L_T = 18\ \text{H} \left(\text{because } L_T = L_1 + L_2 + 2M \right.$$
$$L_T = 3H + 5H + 2(5H) = 18H \right)$$

NOTES

NOTES

NOTES

NOTES

NOTES

NOTES

NOTES

NOTES

NOTES

NOTES